The Chemical Formulary

Collection of Commercial Formulas for Making Thousands of Products in Many Fields

VOLUME XXVIII

Editor-in-Chief

H. BENNETT, F.A.I.C.

Director, B.R. Laboratory (Formula Consultants) Miami Beach, Florida 33140

CHEMICAL PUBLISHING COMPANY, INC.

New York 1989

The Chemical Formulary, Volume XXVIII

ISBN: 978-0-8206-0324-7

Chemical Publishing Company:
www.chemical-publishing.com
www.chemicalpublishing.net

First Edition:

© Chemical Publishing Company, Inc. - New York 1989-2011

Second Impression:

Chemical Publishing Company, Inc. - 2011

Printed in the United States of America

CONTRIBUTORS

Alcolac
3440 Fairfield Rd.
Baltimore, MD 21226

Amerchol Corp.
PO Box 4051, 136 Talmadge Rd.
Edison, NJ 08818

Amoco Chemicals Corp.
200 E. Randolph Dr.
Chicago, IL 60601

Bernel Chemical Co. Inc.
PO Box 777
Tenafly, NJ 07670

Colorcon, Inc.
Moyer Blvd.
West Point, PA 19486

Croda Inc.
51 Madison Ave.
New York, NY 10010

Cyclo Chemical Corp.
7500 N.W. 66th St.
Miami, FL 33166

Dow Chemical Co.
1703 S. Saginaw Rd.
Midland, MI 48640

E.I. DuPont de Nemours & Co.
Nemours Bldg.
Wilmington, DE 19898

Dura Commodities Corp./Durachem
111 Calvert St., PO Box 618
Harrison, NY 10528

FMC Corp.
2000 Market St.
Philadelphia, PA 19103

GAF Corp.
1361 Alps Rd.
Wayne, NJ 07470

General Electric Co.
Silicone Product Div.
Rubber & Fluid Product Dept.
Mechanicsville Rd.
Waterford, NY 12188

Gist-Brocades USA Inc.
PO Box 241068
Charlotte, NC 28224-1068

Goodyear Tire & Rubber Co.
1144 East Market St.
Akron, OH 44316

W.R. Grace & Co./
Evans Chemetics
55 Hayden Ave.
Lexington, MA 02173

Henkel Corp.
300 Brookside Ave.
Ambler, PA 19002

Henkel Resins Div.
4620 W. 77th St.
Minneapolis, MN 55435

ICI Americas
New Murphy Rd. & Concord Pike
Wilmington, DE 19897

Inolex Chem. Co.
Jackson & Swanson Sts.
Philadelphia, PA 19148

Jordan Chemical Co.
1830 Columbia Ave.
Folcroft, PA 19032

Kaopolite Inc.
Office Park
Route 22E, Box 3110
Union, NJ 07083

Lowenstein Dyes & Cosmetics, Inc.
420 Morgan Ave.
Brooklyn, NY 11222

Edward Mendell Co., Inc.
Route 52
Carmel, NY 10512

Mini Fibers Inc.
PO Box 2186
Weber City, VA 24251

Mona Industries Inc.
PO Box 425, 76 E. 24th St.
Paterson, NJ 07544

Nyco
Mt. View Dr.
Willsboro, NY 12996-0368

Patco Products/
Div. C.J. Patterson Co.
3947 Broadway
Kansas City, MO 64111

PPG Industries
One Gateway Center
Pittsburgh, PA 15222

Frank B. Ross Co., Inc.
22 Halladay St., PO Box 4086
Jersey City, NJ 07304-0055

Shell Chemical Co.
One Shell Plaza
Houston, TX 77002

Texaco Chemical Co.
PO Box 15730
Austin, TX 78761

Van Dyk & Co., Inc.
11 Williams St.
Belleville, NJ 07109

R.T. Vanderbilt & Co., Inc.
30 Winfield St.
Norwalk, CT 06855

CONTENTS

PREFACE

Chemistry, as taught in our schools and colleges, concerns chiefly synthesis, analysis, and engineering—and properly so. It is part of the right foundation for the education of the chemist.

Many a chemist entering an industry soon finds that most of the products manufactured by his concern are not synthetic or definite chemical compounds, but are mixtures, blends, or highly complex compounds of which he knows little or nothing. The literature in this field, if any, may be meager, scattered, or obsolete.

Even chemists with years of experience in one or more industries spend considerable time and effort in acquainting themselves with any new field which they may enter. Consulting chemists similarly have to solve problems brought to them from industries foreign to them. There was a definite need for an up-to-date compilation of formulas for chemical compounding and treatment. Since the fields to be covered are many and varied, an editorial board of chemists and engineers engaged in many industries was formed.

Many publications, laboratories, manufacturing firms, and individuals have been consulted to obtain the latest and best information. It is felt that the formulas given in this volume will save chemists and allied workers much time and effort.

Manufacturers and sellers of chemicals will find, in these formulas, new uses for their products. Nonchemical executives, professional men, and interested laymen will make through this volume a "speaking acquaintance" with products which they may be using, trying, or selling.

It often happens that two individuals using the same ingredients in the same formula get different results. This may be due to slight deviations in the raw materials or unfamiliarity with the intricacies of a new technique. Accordingly, repeated experiments may be necessary to get the best results. Although many of the formulas given are being used commercially, many have been taken from the literature and may be subject to various errors and omissions. This should be taken into consideration. Wherever possible, it is advisable to consult with other chemists or technical workers regarding commercial production. This will save time and money and help avoid trouble.

A formula will seldom give exactly the results which one requires. Formulas are useful as starting points from which to work out one's ideas. Also, formulas very often give us ideas which may help us in our specific problems. In a compilation of this kind, errors of omission, commission, and printing may occur. I shall be glad to receive any constructive criticism.

H. BENNETT

PREFACE TO VOLUME XXVIII

This new volume of the CHEMICAL FORMULARY series is a collection of new, up-to-date formulas. The only repetitious material is the introduction (Chapter I) which is used in every volume for the benefit of those who may have bought only one volume and who have no educational background or experience in chemical compounding. The simple basic formulas and compounding methods given in the introduction will serve as a guide for beginners and students. It is suggested that they read the introduction carefully and even make a few preparations described there before compounding the more intricate formulas included in the later chapters.

The list of chemicals and their suppliers has been enlarged with new trademark chemicals. All tradename chemicals appear in the formulas in boldface and these tradenames are listed alphabetically in the appendix followed by a list of corresponding manufacturers. Thus buying the required ingredients will present no problem.

Grateful acknowledgement is made to the Contributors for their valuable suggestions and contributions.

H. BENNETT

NOTE: All the formulas in Volumes I through XXVIII (except in the Introduction) are different. Thus, if you do not find what you want in this volume, you may find it in one of the others.

NOTE: This book is the result of cooperation of many chemists and engineers who have given freely of their time and knowledge. It is their business to act as consultants and to give advice on technical matters for a fee. As publishers, we do not maintain a laboratory or consulting service to compete with them. Therefore, please do not ask us for advice or opinions, but consult a chemist.

BOOKS BY H. BENNETT

The Chemical Formulary Vols. I—XXVII
The Cumulative Index Vols. I–XXV—The Chemical Formulary
Concise Chemical & Technical Dictionary
New Cosmetic Formulary
Chemical Specialties
Industrial Waxes, Vols. I, II
Practical Emulsions, Vols. I, II
More For Your Money
Encyclopedia of Chemical Trademarks and Synonyms, Vols. I–III

ABBREVIATIONS

act.	active, activity
AM	active matter
amp	ampere(s)
anhyd	anhydrous
approx.	approximately
aq.	aqueous
ASTM	American Society for Testing and Materials
avoir	avoirdupois
Bé	Baumé
B.P.	boiling point
°C	degrees Centigrade
cc	cubic centimeter(s)
cm	centimeter(s)
cm^3	cubic centimeter(s)
conc	concentrated, concentration
c.p.	chemically pure
cp, cP, cps	centipoise(s)
cs	centistoke(s)
cu ft	cubic foot
cu in.	cubic inch
cwt	hundredweight
dil	dilute
dm	decimeter(s)
dr	dram(s)
°F	degrees Fahrenheit
fl	fluid
fl dr	fluid dram(s)
fl oz	fluid ounce(s)
F.P.	freezing point
ft	foot
ft^2	square foot
ft^3	cubic foot
g	gram(s)
gal	gallon(s)
gr	grain(s)
h	hour(s)
hl	hectoliter(s)
in.	inch(es)
$in.^3$	cubic inch(es)
K.U.	Krebs units
kg	kilogram(s)
l	liter(s)
lb	pound(s)
liq	liquid

ABBREVIATIONS

m .... milli or meter(s)
MIL .... Military specifications
min .... minimum, minute
ml .... milliliter(s)
mm .... millimeter(s)
M.P. .... melting point
MPa .... mega pascal
N .... newton or Normal
N.F. .... National Formulary
NV .... nonvolatiles
o/w .... oil-in-water
oz .... ounce(s)
P/B .... pigment/binder
Pa .... pascal
pH .... hydrogen-ion concentration
phr .... parts per hundred rubber
pkg. .... package
ppm .... parts per million
psi .... pounds per square inch
psig .... pounds per square inch gauge
pt .... pint(s)
PVC .... pigment volume concentration
pwt .... pennyweight
q.s. .... a quantity sufficient to make
qt .... quart
® .... registered trademark
R.T. .... room temperature
rpm .... revolutions per minute
s .... second(s)
sol'n. .... solution
Sp. Gr. .... specific gravity
T.P. .... triple pressed
tbsp. .... tablespoon(s)
tech .... technical
tinc .... tincture
™ .... trademark
tsp .... teaspoon(s)
USP .... United States Pharmacopeia
UV .... ultraviolet
V .... volt(s)
visc .... viscosity
vol .... volume
w/o .... water-in-oil
wt .... weight

Chapter I

INTRODUCTION

The following introductory matter has been included at the suggestion of teachers of chemistry and home economics.

This section will enable anyone, with or without technical education or experience, to start making simple products without any complicated or expensive machinery. For commercial production, however, suitable equipment is necessary

Chemical specialties are composed of pigments, gums, resins, solvents, oils, greases, fats, waxes, emulsifying agents, dyestuffs, perfumes, water, and chemicals of great diversity. To compound certain of these with some of the others requires definite and well studied procedures, any departure from which will inevitably result in failure. The steps for successful compounding are given with the formulas. Follow them rigorously. If the directions require that (*a*) is added to (*b*), carry this out literally, and do not reverse the order. The preparation of an emulsion is often quite as tricky as the making of mayonnaise. In making mayonnaise, you add the oil to the egg, slowly, with constant and even stirring. If you do it correctly, you get mayonnaise. If you depart from any of these details: if you add the egg to the oil, or pour the oil in too quickly, or fail to stir regularly, the result is a complete disappointment. The same disappointment may be expected if the prescribed procedure of any other formulation is violated.

The point next in importance is the scrupulous use of the proper ingredients. Substitutions are sure to result in inferior quality, if not in complete failure. Use what the formula calls for. If a cheaper product is desired, do not prepare it by substituting a cheaper ingredient for the one prescribed: use a different formula. Not infrequently, a formula will call for an ingredient which is difficult to obtain. In such cases, either reject the formula or substitute a similar substance only after a preliminary experiment demonstrates its usability. There is a limit to which this rule may reasonably be extended. In some cases, substitution of an equivalent ingredient may be made legitimately. For example, when the formula calls for white wax (beeswax), yellow wax can be used, if the color of the finished product is a matter of secondary importance. Yellow beeswax can often replace white beeswax making due allowance for color, but paraffin wax will not replace beeswax, even though its light color seems to place it above yellow beeswax.

This leads to the third point: the use of good-quality ingredients, and ingredients of the correct quality. Ordinary lanolin is not the same thing as anhydrous lanolin. The replacement of one with the other, weight for weight, will give discouragingly different results. Use exactly what the formula calls for: if you are not acquainted with the substance and you are in doubt as to just what is meant, discard the formula and use one you understand. Buy your chemicals from reliable sources. Many ingredients are obtainable in a number of different grades: if the formula does not designate the grade, it is understood that the best grade is to be used. Remember that a formula and the directions can tell you only part of the story. Some skill is often required to attain success. Practice with a small batch in such cases until you are sure of your technique. Many examples can be cited. If the formula calls for steeping quince seed for 30 min in cold water, steeping for 1 h may yield a mucilage of too thin a consistency. The originator of the formula may have used a fresher grade of seed, or his conception of what "cold" water means may be different from yours. You should have a feeling for the right degree of mucilaginousness, and if steeping the seed for 30 min fails to produce it, steep them longer until you get the right kind of mucilage. If you do not know what the right kind is, you will have to experiment until you find out. This is the reason for the recommendation to make small experimental batches until successful results are obtained. Another case is the use of dyestuffs for coloring lotions and the like. Dyes vary in strength; they are all very powerful in tinting value; it is not always easy to state in quantitative terms how much to use. You must establish the quantity by carefully adding minute quantities until you have the desired tint. Gum tragacanth is one of those products which can give much trouble. It varies widely in solubility and bodying power; the quantity listed in the formula may be entirely unsuitable for your grade of tragacanth. Therefore, correction is necessary, which can be made only after experiments with the available gum.

In short, if you are completely inexperienced, you can profit greatly by experimenting. Such products as mouthwashes, hair tonics, and astringent lotions need little or no experience, because they are, as a rule, merely mixtures of simple liquid and solid ingredients, which dissolve without difficulty and the end product is a clear solution that is ready for use when mixed. However, face creams, toothpastes, lubricating greases, wax polishes, etc., whose formulation requires relatively elaborate procedures and which must have a definite final viscosity, need some skill and not infrequently some experience.

FIGURING

Some prefer proportions expressed by weight or volume, others use percentages. In different industries and foreign countries different systems of weights and measures are used. For this reason, no one set of units could be satisfactory for everyone. Thus diverse formulas appear with different units, in accordance with their sources of origin. In some cases, parts are given instead of percentage or weight or volume. On the pages preceding the index, conversion tables of weights and measures are listed. These are used for changing from one system to another. The following examples illustrate typical units.

Example No. 1

Ink for Marking Glass

Glycerin	40	Ammonium Sulfate	10
Barium Sulfate	15	Oxalic Acid	8
Ammonium Bifluoride	15	Water	12

Here no units are mentioned. In this case, it is standard practice to use parts by weight throughout. Thus here we may use ounces, grams, pounds, or kilograms as desired. But if ounces are used for one item, the ounce must be the unit for all the other items in the formula.

Example No. 2

Flexible Glue

Powdered Glue	30.90%	Glycerin	5.15%
Sorbitol (85%)	15.45%	Water	48.50%

Where no units of weight or volume, but percentages are given, forget the percentages and use the same method as given in Example No. 1.

Example No. 3

Antiseptic Ointment

Petrolatum	16 parts	Benzoic Acid	1 part
Coconut Oil	12 parts	Chlorothymol	1 part
Salicylic Acid	1 parts		

The instructions given for Example No. 1 also apply to Example No. 3. In many cases, it is not wise to make up too large a quantity of a product before making a number of small batches to first master the necessary technique and also to see whether the product is suitable for the particular purpose for which it is intended. Since, in many cases, a formula may be given in proportions as made up on a factory scale, it is advisable to reduce the quantities proportionately.

Example No. 4

Neutral Cleansing Cream

Mineral Oil	80 lb	Water	90 lb
Spermaceti	30 lb	Glycerin	10 lb
Glyceryl Monostearate	24 lb	Perfume	To suit

Here, instead of pounds, ounces or even grams may be used. This formula would then read:

Mineral Oil	80 g	Water	90 g
Spermaceti	30 g	Glycerin	10 g
Glyceryl Monostearate	24 g	Perfume	To suit

Reduction in bulk may also be obtained by taking the same fractional part of portion of each ingredient in a formula. Thus in the following formula:

Example No. 5

Vinegar Face Lotion

Acetic Acid (80%)	20	Alcohol	440
Glycerin	20	Water	500
Perfume	20		

We can divide each amount by ten and then the finished bulk will be only one tenth of the original formula. Thus it becomes:

Acetic Acid (80%)	2	Alcohol	44
Glycerin	2	Water	50
Perfume	2		

APPARATUS

For most preparations, pots, pans, china, and glassware, which are used in every household, will be satisfactory. For making fine mixtures and emulsions, a malted-milk mixer or egg beater is necessary. For weighing, a small, low-priced scale should be purchased from a laboratory-supply house. For measuring fluids, glass graduates or measuring glasses may be purchased from your local druggist. Where a thermometer is necessary, a chemical thermometer should be obtained from a druggist or chemical-supply firm.

METHODS

To understand better the products which you intend to make, it is advisable that you read the complete section covering such products. You may learn different methods that may be used and also to avoid errors which many beginners are prone to make.

CONTAINERS FOR COMPOUNDING

Where discoloration or contamination is to be avoided, as in light-colored, or food and drug products, it is best to use enameled or earthenware vessels. Aluminum is also highly desirable in such cases, but it should not be used with alkalis as these dissolve and corrode aluminum.

HEATING

To avoid overheating, it is advisable to use a double boiler when temperatures below 212 F (temperature of boiling water) will suffice. If a double boiler is not at hand, any pot may be filled with water and the vessel containing the ingredients to be heated placed in the water. The pot may then be heated by any flame without fear of overheating. The water in the pot, however, should be replenished from time to time; it must not be allowed to "go dry." To get uniform higher temperatures, oil, grease, or wax is used in the outer container in place of water. Here, of course, care must be taken to stop heating when thick fumes are given off as these are inflammable. When higher uniform temperatures are necessary, molten lead may be used as a heating medium. Of course, with chemicals which melt uniformly and are nonexplosive, direct heating over an open flame is permissible, with stirring, if necessary.

Where instructions indicate working at a certain temperature, it is important to attain the proper temperature not by guesswork, but by the use of a thermometer. Deviations from indicated temperatures will usually result in spoiled preparations.

TEMPERATURE MEASUREMENT

In the United States and in Great Britain, the Fahrenheit scale of temperature is used. The temperature of boiling water is 212 Fahrenheit (212 F); the temperature of melting ice is 32 Fahrenheit (32 F).

In scientific work, and in most foreign countries, the Centigrade scale is used, on which the temperature of boiling water is 100 Centigrade (100 C) and the temperature of melting ice is 0 Centigrade (0 C).

The temperature of liquids is measured by a glass thermometer. This is inserted as deeply as possible in the liquid and is moved about until the temperature reading remains steady. It takes a short time for the glass of the thermometer to reach the temperature of the liquid. The thermometer should not be placed against the bottom or side of the container, but near the center of the liquid in the vessel. Since the glass of the thermometer bulb is very thin, it breaks easily when striking it against any hard surface. A cold thermometer should be warmed gradually (by holding it over the surface of a hot liquid) before immersion. Similarly the hot thermometer when taken out of the liquid should not be put into cold water suddenly. A sharp change in temperature will often crack the glass.

MIXING AND DISSOLVING

Ordinary dissolution (e.g., that of sugar in water) is hastened by stirring and warming. Where the ingredients are not corrosive, a clean stick, a fork, or spoon may be used as a stirring rod. These may also be used for mixing thick creams or pastes. In cases where very thorough stirring is necessary (e.g., in making mayonnaise, milky polishes, etc.), an egg beater or a malted-milk mixer is necessary.

FILTERING AND CLARIFICATION

When dirt or undissolved particles are present in a liquid, they are removed by settling or filtering. In the first procedure, the solution is allowed to stand and if the particles are heavier than the liquid they will gradually sink to the bottom. The liquid may be poured or siphoned off carefully and, in some cases, it is then sufficiently clear for use. If, however, the particles do not settle out, then they must be filtered off. If the particles are coarse they may be filtered or strained through muslin or other cloth. If they are very small, filter paper is used. Filter papers may be obtained in various degrees of fineness. Coarse filter paper filters rapidly but will not retain extremely fine particles. For fine particles, a very fine grade of filter paper should be used. In extreme cases, even this paper may not be fine enough. Then it will be necessary to add to the liquid 1–3% infusorial earth or magnesium carbonate. These are filter aids that clog up the pores of the filter paper and thus reduce their size and hold back undissolved material of extreme fineness. In all such filtering, it is advisable to take the first portions of the filtered liquid and pour them through the filter again as they may develop cloudiness on standing.

DECOLORIZING

The most commonly used decolorizer is decolorizing carbon. This is added to the liquid to the extent of 1–5% and the liquid is heated, with stirring, for $^1/_2$ hour to as high a temperature as is feasible. The mixture is then allowed to stand for a while and filtered. In some cases, bleaching must be resorted to.

PULVERIZING AND GRINDING

Large masses or lumps are first broken up by wrapping in a clean cloth, placing between two boards, and pounding with a hammer. The smaller pieces are then pounded again to reduce their size. Finer grinding is done in a mortar with a pestle.

SPOILAGE AND LOSS

All containers should be closed when not in use to prevent evaporation or contamination by dust; also because, in some cases, air affects the product adversely. Many chemicals attack or corrode the metal containers in which they are kept. This is particularly true of liquids. Therefore, liquids should be transferred into glass bottles which should be as full as possible. Corks should be covered with aluminum foil (or dipped in melted paraffin wax when alkalies are present).

Glue, gums, olive oil, or other vegetable or animal products may ferment or become rancid. This produces discoloration or unpleasant odors. To avoid this, suitable antiseptics or preservatives must be used. Cleanliness is of utmost importance. All containers must be cleaned thoroughly before use to avoid complications.

WEIGHING AND MEASURING

Since, in most cases, small quantities are to be weighed, it is necessary to get a light scale. Heavy scales should not be used for weighing small amounts as they are not accurate enough for this type of weighing.

For measuring volumes of liquids, measuring glasses or cylinders (graduates) should be used. Since this glassware cracks when heated or cooled suddenly it should not be subjected to sudden changes of temperature.

CAUTION

Some chemicals are corrosive and poisonous. In many they are labeled as such. As a precautionary measure, it is advised not to inhale them and, if smelling is absolutely necessary, only to sniff a few inches from the cork or stopper. Always work in a well-ventilated room when handling poisonous or unknown chemicals. If anything is spilled, it should be wiped off and washed away at once.

WHERE TO BUY CHEMICALS AND APPARATUS

Many chemicals and most glassware can be purchased from your druggist. A list of suppliers of all products is at the end of this book.

ADVICE

This book is the result of cooperation of many chemists and engineers who have given freely of their time and knowledge. It is their business to act as consultants and to give advice on technical matters for a fee. As publishers, we do not maintain a laboratory or consulting service to compete with them.

Please, therefore, do not ask us for advice or opinions, but confer with a chemist in your vicinity.

EXTRA READING

Keep up with new developments of materials and methods by reading technical magazines. Many technical publications are listed under references in the back of this book.

CALCULATING COSTS

Raw materials, purchased in small quantities, are naturally higher in price than when bought in large quantities. Commercial prices, as given in the trade papers and catalogs of manufacturers, are for large quantities such as barrels, drums, or sacks. For example, 1 lb epsom salts, bought at retail, may cost 10–15 cents. In barrel lots its price is much lower.

Typical Costing Calculation Formula for Beer- or Milk-Pipe Cleaner

Soda Ash	25 lb	@ $\$0.02\frac{1}{2}$ per lb =	$ 0.63
Sodium Perborate	75 lb	@ 0.16 per lb =	12.00
	Total 100 lb		$12.63

If 100 lb cost $12.63, 1 lb will cost $12.63 divided by 100 or about $0.126, assuming no loss.

Always weigh the amount of finished product and use this weight in calculating costs. Most compounding results in some loss of material because of spillage, sticking to apparatus, evaporation, etc. Costs of making experimental lots are always high and should not be used for figuring costs. To meet competition, it is necessary to buy in large quantities and manufacturing costs should be based on these.

ELEMENTARY PREPARATIONS

The simple formulas that follow have been selected because of their importance and because they are easy to make.

The succeeding chapters go into greater detail and give many different types and modifications of these and other recipes for home and commercial use.

Cleansing Creams

Cleansing creams, as the name implies, serve as skin cleaners. Their basic ingredients are oils and waxes which are rubbed into the skin. When wiped off, they carry off dirt and dead skin. The liquefying type cleansing cream contains no high-melting matter and melts or liquefies when rubbed on the skin. To suit different climates and likes and dislikes harder or softer products can be made.

Cleansing Cream (Liquefying)

Liquid Petrolatum	5.5
Paraffin Wax	2.5
Petrolatum	2.0

Melt the ingredients together, with stirring, in an aluminum or enameled dish and allow to cool. Then stir in a perfume oil. Allow to stand until it becomes waxy and then pour into jars, which should be allowed to stand undisturbed overnight.

Cold Creams

The most important facial cream is the cold cream. This type of cream contains mineral oil and wax which are emulsified in water with a small amount of borax or glyceryl monostearate, S.E. The function of a cold cream is to form a film that takes up dirt and waste tissue, which are removed when the skin is wiped thoroughly. Many modifications of this basic cream are encountered in stores. They vary in color, odor, and in claims, but, essentially, they are not more useful than this simple cream. The latest type of cold cream is the nongreasy cold cream which is of particular interest because it is nonalkaline and, therefore, nonirritating for sensitive skins.

Cold Cream

Liquid Petrolatum	52 g
White Beeswax	14 g

Heat this in an aluminum or enameled double boiler. (The water in the outer pot should be brought to a boil.) In a separate aluminum or enameled pot dissolve:

Borax	1 g
Water	33 cc

and bring this to a boil. Add this in a thin stream to the melted wax, while stirring vigorously in one direction only. When the temperature drops to 140 F, add 0.5 cc perfume oil and continue stirring until the temperature drops to 120 F. At this point, pour into jars, where the cream will set after a while. If a harder cream is desired, reduce the amount of liquid petrolatum. If a softer cream is wanted, increase it.

Nongreasy Cold Cream

White Paraffin Wax	1.25
Petrolatum	1.50
Glyceryl Monostearate, S.E.	2.25
Liquid Petrolatum	3.00

Heat this mixture in an aluminum or enameled double boiler. (The water in the outer pot should be boiling.) Stir until clear. To this slowly add, while stirring vigorously:

Boiling Water	10

Continue stirring until smooth and then add, with stirring, perfume oil. Pour into jars at 110–130 F and cover the jars as soon as possible.

Vanishing Creams

Vanishing creams are nongreasy soapy creams which have a cleansing effect. They are also used as a powder base.

Vanishing Cream

Stearic Acid	18 oz

Melt this in an aluminum or enameled double boiler. (The water in the outer pot must be boiling.) Add, in a thin stream, while stirring vigorously, the following boiling solution made in an aluminum or enameled pot:

Potassium Carbonate	$^1/_4$ oz
Glycerin	$6^1/_2$ oz
Water	5 lb

Continue stirring until the temperature falls to 135 F, then mix in a perfume oil and stir from time to time until cold. Allow to stand overnight and stir again the next day. Pack into jars and close these tightly.

Hand Lotions

Hand lotions are usually clear or milky liquids or salves which are useful in protecting the skin from roughness and redness because of exposure to cold, hot water, soap, and other agents. Chapped hands are common. The use of a good hand lotion keeps the skin smooth, soft, and in a healthy condition. The lotion is best applied at night, rather freely, and cotton gloves may be worn to prevent soiling. During the day, it should be put on sparingly and the excess wiped off.

Hand Lotion (Salve)

Boric Acid	1
Glycerin	6

Warm these in an aluminum or enameled dish and stir until dissolved (clear). Then allow to cool and work this liquid into the following mixture, adding only a little at a time.

Lanolin	6
Petrolatum	8

To impart a pleasant odor a little perfume may be added and worked in.

Hand Lotion (Milky Liquid)

Lanolin	1/4 tsp
Glyceryl Monostearate, S.E.	1 oz
Tincture of Benzoin	2 oz
Witch Hazel	25 oz

Melt the first two items together in an aluminum or enameled double boiler. If no double boiler is at hand, improvise one by placing a dish in a small pot containing boiling water. When the mixture becomes clear, remove from the double boiler and add slowly, while stirring vigorously, the tincture of benzoin and then the witch hazel. Continue stirring until cool and then put into one or two large bottles and shake vigorously. The finished lotion is a milky liquid comparable to the best hand lotions on the market sold at high prices.

Brushless Shaving Creams

Brushless or latherless shaving creams are soapy in nature and do not require lathering or water. The formula given here is of the latest type being free from alkali and nonirritating. It should be borne in mind, however, that certain beards are not softened by this type of cream and require the old fashioned lathering shaving cream.

Brushless Shaving Cream

White Mineral Oil	10
Glyceryl Monostearate, S.E.	10
Water	50

Heat the first two ingredients together in a Pyrex or enameled dish to 150 F and run in slowly, while stirring, the water which has been heated to boiling. Allow to cool to 150 F and, while stirring, add a few drops of perfume oil. Continue stirring until cold.

Mouthwashes

Mouthwashes and oral antiseptics are of practically negligible value. However, they are used because of their refreshing taste and slight deodorizing effect.

Mouthwash

Benzoic Acid	$^{5}/_{8}$
Tincture of Rhatany	3
Alcohol	20
Peppermint Oil	$^{1}/_{8}$

Mix together in a dry bottle until the benzoic acid is dissolved. One teaspoonful is used to a small wine-glassful of water.

Tooth Powders

The cleansing action of tooth powders depends on their contents of soap and mild abrasives, such as precipitated chalk and magnesium carbonate. The antiseptic present is practically of no value. The flavoring ingredients mask the taste of the soap and give the mouth a pleasant aftertaste.

Tooth Powder

Magnesium Carbonate	420 g
Precipitated Chalk	565 g
Sodium Perborate	55 g
Sodium Bicarbonate	45 g
Powdered White Soap	50 g
Powdered Sugar	90 g
Wintergreen Oil	8 cc
Cinnamon Oil	2 cc
Menthol	1 g

Dissolve the last three ingredients together and then rub well into the sugar. Add the soap and perborate, mixing well. Add the chalk, with good mixing, and then the sodium bicarbonate and magnesium carbonate. Mix thoroughly and sift through a fine wire screen. Keep dry.

Foot Powders

Foot powders consist of talc or starch with or without an antiseptic or deodorizer. In the following formula the perborates liberate oxygen, when in contact with perspiration, which tends to destroy unpleasant odors. The talc acts as a lubricant and prevents friction and chafing.

Foot Powder

Sodium Perborate	3
Zinc Peroxide	2
Talc	15

Mix thoroughly in a dry container until uniform. This powder must be kept dry or will spoil.

Liniments

Liniments usually consist of an oil and an irritant, such as methyl salicylate or turpentine. The oil acts as a solvent and tempering agent for the irritant. The irritant produces a rush of blood and warmth which is often slightly helpful.

Sore-Muscle Liniment

Olive Oil	6 fl oz
Methyl Salicylate	3 fl oz

Mix together and keep in a well-stoppered bottle. Apply externally, but do not use on chafed or cut skin.

Chest Rubs

In spite of the fact that chest rubs are practically useless, countless sufferers use them. Their action is similar to that of liniments and they differ only in that they are in the form of a salve

Chest-Rub Salve

Yellow Petrolatum	1 lb
Paraffin Wax	1 oz
Eucalyptus Oil	2 fl oz
Menthol	1/2 oz
Cassia Oil	1/8 fl oz
Turpentine	1/2 fl oz

Melt the petrolatum and paraffin wax together in a double boiler and then add the menthol. Remove from the heat, stir, and cool a little; then mix in the oils, and turpentine. When it begins to thicken, pour into tins and cover.

Insect Repellents

Preparations of this type may irritate sensitive skin and they will not always work.

Mosquito-Repelling Oil

Cedar Oil	2 fl oz
Citronella Oil	4 fl oz
Spirits of Camphor	8 fl oz

Mix in a dry bottle and the oil is ready for use. This preparation may be smeared on the skin as often as is necessary.

Fly Sprays

Fly sprays usually consist of deodorized kerosene, perfume, and an active insecticide. In some cases, they merely stun the flies who may later recover and begin buzzing again.

Fly Spray

Deodorized Kerosene	80 fl oz
Methyl Salicylate	1 fl oz
Pyrethrum Powder	10 oz

Mix thoroughly by stirring from time to time; allow to stand covered overnight and then filter through muslin.

This spray is inflammable and should not be used near open flames.

Disinfectants

Deodorant Spray
(For public buildings, sick rooms, lavatories, etc.)

Pine Needle Oil	2
Formaldehyde	2
Acetone	6
* Isopropyl Alcohol	20

1 oz of this mixture is diluted with 1 pt water for spraying.

Cresol

† Caustic Soda	25.5 g
Water	140.0 cc

Dissolve in a Pyrex or enameled dish and warm. To this, add slowly the following warmed mixture:

** Cresylic Acid	500.0 cc
Rosin	170.0 g

Stir until dissolved and add water to make 1000 cc.

---

* Inflammable.

** Poison.

† Do not get this on the skin as it is corrosive.

Insecticides

Ant Poison

	Sugar	1 lb
	Water	1 qt
**	Arsenate of Soda	125 g

Boil and stir until uniform; strain through muslin and add 1 spoonful honey.

Bedbug Exterminator

*	Kerosene	90 fl oz
	Clove Oil	5 fl oz
**	Cresol	1 fl oz
	Pine Oil	4 fl oz

Simply mix and bottle.

Nonstaining Mothproofing Fluid

Sodium Aluminum Silicofluoride	0.50
Water	98.00
Glycerin	0.50
"Sulfatate" (Wetting Agent)	0.25

Stir until dissolved.

Fly Paper

Rosin	32
Rosin Oil	20
Castor Oil	8

Heat this mixture in an aluminum or enameled pot on a gas stove, with stirring, until all the rosin has melted and dissolved. While hot, pour on firm paper sheets of suitable size which have been brushed with soap water just

* Inflammable.
** Poison.

before coating. Smooth out the coating with a long knive or piece of thin flat wood and allow to cool. If a heavier coating is desired, increase the amount of rosin. Similarly, a thinner coating results by reducing the amount of rosin. The finished paper should be laid flat and not exposed to undue heat.

Baking Powder

Bicarbonate of Soda	28
Monocalcium Phosphate	35
Corn Starch	27

Mix these powders thoroughly in a dry can by shaking and rolling for 1/2 hour. Pack into dry airtight tins as moisture will cause lumping.

Malted-Milk Powder

Powdered Malt Extract	5
Powdered Skim Milk	2
Powdered Sugar	3

Mix thoroughly by shaking and rolling in a dry can. Pack in an airtight container.

Cocoa-Malt Powder

Corn Sugar	55
Powdered Malt	19
Powdered Skim Milk	12 1/2
Cocoa	13
Vanillin	1/8
Powdered Salt	3/8

Mix thoroughly and then run through a fine wire sieve.

Sweet Cocoa Powder

Cocoa	17 1/2 oz
Powdered Sugar	32 1/2 oz
Vanillin	3/4 g

Mix thoroughly and sift.

Pure Lemon Extract

Lemon Oil U.S.P.	$6\frac{1}{2}$ fl oz
Alcohol	$121\frac{1}{2}$ fl oz

Shake together in 1-gal jug until dissolved.

Artificial Vanilla Flavor

Vanillin	$\frac{3}{4}$ oz
Alcohol	2 pt

Stir the ingredients in a glass or china pitcher until dissolved. Then mix into the following solution:

Sugar	12 oz
Water	$5\frac{1}{4}$ pt
Glycerin	1 pt

Color brown by adding sufficient burnt-sugar coloring.

Canary Food

Dried and Chopped Egg Yolk	2
Poppy Heads (coarse powder)	1
Cuttlefish Bone (coarse powder)	1
Powdered Soda Crackers	8

Mix well together.

Blue-Black Writing Ink

Naphthol Blue-Black	1 oz
Powdered Gum Arabic	$\frac{1}{2}$ oz
Carbolic Acid	$\frac{1}{4}$ oz
Water	1 gal

Stir together in a glass or enameled vessel until dissolved.

Indelible Laundry-Marking Ink

A	Soda Ash	1 oz
	Powdered Gum Arabic	1 oz
	Water	10 fl oz

Stir until dissolved.

B	Silver Nitrate	4 oz
	Powdered Gum Arabic	4 oz
	Lampblack	2 oz
	Water	40 fl oz

Stir this in a glass or porcelain dish until dissolved. Do not expose the mixture to strong light or it will spoil. Then pour into a brown glass bottle. In using these solutions, wet the cloth with solution A and allow to dry. Then write on it with solution B using a quill pen.

Green Marking Crayon

Ceresin	8
Carnauba Wax	7
Paraffin Wax	4
Beeswax	1
Talc	10
Chrome Green	3

Melt the first four ingredients in a container and then add the last two slowly, while stirring. Remove from the heat and continue stirring until thickening begins. Then pour into molds. If other-color crayons are desired, other pigments may be used. For example, for black, use carbon black or bone black; for blue, Prussian blue; for red, orange chrome yellow.

Antique Coloring for Copper

Copper Nitrate	4 oz
Acetic Acid	1 oz
Water	2 oz

Dissolve by stirring together in a glass or porcelain vessel. Pack into glass bottles.

Wet the copper to be colored and apply the coloring solution hot.

Blue-Black Finish on Steel

A Place the object in molten sodium nitrate at 700–800 F for 2–3 min. Remove and allow to cool somewhat, wash in hot water, dry, and oil with mineral or linseed oil.

B Then put the object in the following solution for 15 min:

Copper Sulfate	1/2 oz
Iron Chloride	1 lb
* Hydrochloric Acid	4 oz
** Nitric Acid	1/2 oz
Water	1 gal

Allow to dry for several hours. Place in a solution again for 15 min, remove and dry for 10 h. Place in boiling water for 1/2 h, dry, and scratch-brush very lightly. Oil with mineral or linseed oil and wipe dry.

Rust-Prevention Compound

Lanolin	1
* Naphtha	2

Mix until dissolved.

The metal to be protected is cleaned with a dry cloth and then coated with the composition.

Metal Polish

Naphtha	62 oz
Oleic Acid	1/3 oz
Abrasive	7 oz
Triethanolamine Oleate	1/3 oz
Ammonia (26%)	1 oz
Water	1 gal

---

* Inflammable.

** Corrosive.

In one container mix together the naphtha and oleic acid to a clear solution. Dissolve the triethanolmaine oleate in the water separately, stir in the abrasive, and then add the naphtha solution. Stir in the resulting mixture at a high speed until a uniform creamy emulsion results. Then add the ammonia and mix well, but do not agitate so vigorously as before.

Glass-Etching Fluid

Hot Water	12
† Ammonium Bifluoride	15
Oxalic Acid	8
Ammonium Sulfate	10
Glycerin	40
Barium Sulfate	15

Warm the washed glass slightly before writing on it with this fluid. Allow the fluid to act on the glass for about 2 min.

This is an excellent preservative for leather book bindings, luggage, and other leather goods.

White-Shoe Dressing

Lithopone	19 oz
Titanium Dioxide	1 oz
Bleached Shellac	3 oz
Ammonium Hydroxide	$^1/_4$ fl oz
Water	25 fl oz
Glycerin	1 oz

Dissolve the last four ingredients by mixing in a porcelain vessel. When dissolved stir in the first two pigments. Keep in stoppered bottle and shake before using.

Waterproofing for Shoes

Wool Grease	8
Dark Petrolatum	4
Paraffin Wax	4

Melt together in any container.

† Corrosive.

Leather Preservative
Cold-Pressed Neatsfoot

Mineral Oil	10
Castor Oil	10

Mix.

Polishes

Polishes are generally used to restore the original luster and finish of a smooth surface. They are also expected to clean the surface and to prevent corrosion or deterioration. There is no one polish which will give good results on all surfaces.

Most polishes contain oil or wax for their lustering or polishing properties. Oil polishes are easy to apply, but the surfaces on which they are used attract dust and show finger marks. Wax polishes are more difficult to apply, but are more lasting.

Oil or wax polishes are of two types: waterless and aqueous. The former are clear or translucent, the latter are milky in appearance.

For use on metals, abrasives of various kinds, such as tripoli, silica dust, or infusorial earth, are incorporated to grind away oxide films or corrosion products.

Black Shoe Polish

Carnauba Wax	$5^1/_2$ oz
Crude Montan Wax	$5^1/_2$ oz

Melt together in a double boiler. (The water in the outer container should be boiling.) Then stir in the following metled and dissolved mixture:

Stearic Acid	2 oz
Nigrosine Base	1 oz

Then stir in

Ceresin	15 oz

Remove all flames and run in slowly, while stirring.

Turpentine	90 fl oz

Allow the mixture to cool to 105 F and pour into airtight tins which should stand undisturbed overnight.

Clear Oil-Type Auto Polish

Paraffin Oil	5 pt
Raw Linseed Oil	2 pt
China-Wood Oil	1/2 pt
* Benzol	1/4 pt
* Kerosene	1/4 pt
Amyl Acetate	1 tbsp

Mix together in a glass jar and keep it stoppered.

Paste-Type Auto and Floor Wax

Yellow Beeswax	1 oz
Ceresin	2 1/2 oz
Carnauba Wax	4 1/2 oz
Montan Wax	1 1/4 oz
* Naphtha or Mineral Spirits	1 pt
* Turpentine	2 oz
Pine Oil	1/2 oz

Melt the waxes together in a double boiler. Turn off the heat and run in the last three ingredients in a thin stream, with stirring. Pour into cans, cover, and allow to stand undisturbed overnight.

Oil-and-Wax Type Furniture Polish

Paraffin Oil	1 pt
Powdered Carnauba Wax	1/4 oz
Ceresin Wax	1/8 oz

Heat together until all of the wax is melted. Allow to cool and pour into bottles before the mixture turns cloudy.

---

* Inflammable—keep away from flames.

Liquid Polishing Wax

Yellow Beeswax	1 oz
Ceresin Wax	4 oz

Melt together and then cool to 130 F; turn off all flames and stir in slowly:

* Turpentine	17 fl oz
Pine Oil	1/2 fl oz

Pour into cans or bottles which are closed tightly to prevent evaporation.

Floor Oil

Mineral Oil	46 fl oz
Beeswax	1/2 oz
Carnauba Wax	1 oz

Heat together in double boiler until dissolved (clear). Turn off the flame and stir in

* Turpentine	3 fl oz

Lubricants

Lubricants, in the form of oils or greases, are used to prevent friction and wearing of parts which are rubbed together. Lubricants must be chosen to fit specific uses. They consist of oils and fats often compounded with soaps and other unctuous substances. For heavy duty, heavy oils or greases are used and light oils are suitable for light duty.

Gum Lubricant

White Petrolatum	15 oz
Acid-Free Bone Oil	5 oz

Warm gently and mix together.

* Inflammable.

Graphite Grease

Ceresin	7 oz
Tallow	7 oz

Warm together and gradually work in with a stick:

Graphite	3 oz

Stir until uniform and pack in tins when thickening begins.

Penetrating Oil
(For loosening rusted bolts, screws, etc.)

Kerosene	2 oz
Thin Mineral Oil	7 oz
Secondary Butyl Alcohol	1 oz

Mix and keep in a stoppered bottle.

Molding Compound

White Glue	13 lb
Rosin	13 lb
Raw Linseed Oil	1/3 qt
Glycerin	1 qt
Whiting	19 lb

Heat the white glue until it melts. Then cook separately the rosin and raw linseed oil until the rosin is dissolved. Add the rosin, oil, and glycerin to the glue, stirring in the whiting until the mass reaches the consistency of a putty. Keep the mixture hot.

Press this mass into the die firmly and allow it to cool slightly before removing. The finished product is ready to use within a few hours after removal. Suitable pigments may be added to secure brown, red, black, or other color.

In applying ornaments made of this composition to a wood surface, they are first steamed to make them flexible; in this condition, they will adhere to the wood easily and securely. They can be bent to any shape, and no nails are required for applying them.

Grafting Wax

Wool Grease	11
Rosin	22
Paraffin Wax	6
Beeswax	4
Japan Wax	1
Rosin Oil	9
Pine Oil	1

Melt together until clear and pour into tins. This composition can be made thinner by increasing the amount of rosin oil and thicker by decreasing it.

Candles

Paraffin Wax	30.0
Stearic Acid	17.5
Beeswax	2.5

Melt together and stir until clear. If colored candles are desired, add a very small amount of any oil-soluble dye. Pour into vertical molds in which wicks are hung.

Adhesives

Adhesives are sticky substances used to unite two surfaces. Adhesives are specifically called glues, pastes, cements, mucilages, lutes, etc. For different uses different types are required.

Wall-Patching Plaster

Plaster of Paris	32
Dextrin	4
Pumice Powder	4

Mix thoroughly by shaking and rolling in a dry container. Keep away from moisture.

Cement-Floor Hardener

Magnesium Fluosilicate	1 lb
Water	15 pt

Mix until dissolved.

The cement should first be washed with clean water and then drenched with this solution.

Paperhanger's Paste

White or Fish Glue	4 oz
Cold Water	8 oz
Venice Turpentine	2 fl oz
Rye Flour	1 lb
Cold Water	16 fl oz
Boiling Water	64 fl oz

Soak the glue in the first amount of cold water for 4 h. Dissolve on a waterbath (glue-pot) and while hot stir in the Venice turpentine. Use a cheap grade of rye or wheat flour, mix thoroughly with the second amount of cold water to about the consistency of dough or a little thinner, being careful to remove all lumps. Stir in 1 tbsp of powdered alum to 1 qt flour, then pour in the boiling water, stirring rapidly until the flour is thoroughly cooked. Let this cool and finally add the glue solution. This makes a very strong paste which will also adhere to a painted surface, owing to the Venice turpentine content.

Aquarium Cement

Litharge	10
Plaster of Paris	10
Powdered Rosin	1
Dry White Sand	10
Boiled Linseed Oil	Sufficient

Mix all together in the dry state, and make a stiff putty with the oil just before use.

Do not fill the aquarium for 3 days after cementing. This cement hardens under water, and will stick to wood, stone, metal, or glass and as it resists the action of sea water, it is useful for marine *aquaria*.

Wood-Dough Plastic

* Collodion	86
Powdered Ester Gum	9
Wood Flour	30

Allow the first two ingredients to stand until dissolved, stirring from time to time. Then, while stirring, add the wood flour, a little at a time, until uniform. This product can be made softer by adding more collodion.

Putty

Whiting	80
Raw Linseed Oil	16

Rub together until smooth. Keep in a closed container.

Wood-Flour Bleach

Sodium Metasilicate	90
Sodium Perborate	10

Mix thoroughly and keep dry in a closed can. Use 1 lb to 1 gal boiling water. Mop or brush on the floor, allow to stand 1/2 h, then rub off and rinse well with water.

*Paint Remover

Benzol	5 pt
Ethyl Acetate	3 pt
Butyl Acetate	2 pt
Paraffin Wax	1/2 lb

Stir together until dissolved.

* Inflammable.

Soaps and Cleaners

Soaps are made from a fat or fatty acid and an alkali. They lather and produce a foam which entraps dirt and grease. There are many kinds of soaps.

Cleaners contain a solvent, such as naphtha, with or without a soap. Abrasive cleaners are soap pastes containing powdered pumice, stone, silica, etc.

Concentrated Liquid Soap

Water	11
* Solid Caustic Potash	1
Glycerin	4
Red Oil (Oleic Acid)	4

Dissolve the caustic soda in water, add the glycerin, and bring to a boil in an enameled pot. Remove from the heat, add the red oil slowly, while stirring. If a more neutral soap is wanted, use more red oil.

Saddle Soap

Beeswax	5.0
* Caustic Potash	0.8
Water	8.0

Boil for 5 min, while stirring. In another vessel heat:

Castile Soap	1.6
Water	8.0

Mix the two solutions with good stirring; remove from the heat and add, while stirring:

Turpentine	12

* Do not get on the skin as it is corrosive.

Mechanics' Hand-Soap Paste

Water	1.8 qt
White Soap Chips	1.5 lb
Glycerin	2.4 oz
Borax	8.4 oz
Dry Sodium Carbonate	3.0 oz
Coarse Pumice Powder	2.2 lb
Safrol	To suit

Dissolve the soap in two-thirds of the water by heat. Dissolve the last three ingredients in the rest of the water. Pour the two solutions together and stir well. When it begins to thicken, sift in the borax, stirring constantly till thick, then pour into cans. Vary the amount of water, for heavier or softer paste. Water cannot be added to the finished soap.

Dry-Cleaning Fluid

Glycol Oleate	2 fl oz
Carbon Tetrachloride	60 fl oz
Naphtha	20 fl oz
Benzene	18 fl oz

This is an excellent cleaner that will not injure the finest fabrics.

Wall-Paper Cleaner

Whiting	10 lb
Calcined Magnesia	2 lb
Fuller's Earth	2 lb
Powdered Pumice	12 oz
Lemenone or Citronella Oil	4 oz

Mix well together.

Household Cleaner

Soap Powder	2
Soda Ash	3
Trisodium Phosphate	40
Finely Ground Silica	55

Mix well and pack in the usual containers.

Window Cleanser

Castile Soap	2
Water	5
Chalk	4
French Chalk	3
Tripoli Powder	2
Petroleum Spirits	5

Mix well and pack in tight containers.

Straw-Hat Cleaner

Sponge the hat with a solution of:

Sodium Hyposulfite	10 oz
Glycerin	5 oz
Alcohol	10 oz
Water	75 oz

Lay the hat aside in a damp place for 24 h and then apply a mixture of:

Citric Acid	2 oz
Alcohol	10 oz
Water	90 oz

Press with a moderately hot iron after stiffening with gum water, if necessary.

Grease, Oil, Paint, and Lacquer Spot Remover

Alcohol	1
Ethyl Acetate	2
Butyl Acetate	2
Toluol	2
Carbon Tetrachloride	3

Place the garment with the spot over a piece of clean paper or cloth and wet the spot with this fluid; rub with a clean cloth toward the center of the spot. Use a clean section of cloth for rubbing and clean paper or cloth for each application of the fluid. This cleaner is inflammable and should be kept away from flames. Cleaners of this type should be used out of doors or in well ventilated rooms as the fumes are toxic.

Paint-Brush Cleaner

A	Kerosene	2.00
	Oleic Acid	1.00
B	Strong Liquid Ammonia (28%)	0.25
	Denatured Alcohol	0.25

Stir slowly B into A until a smooth mixture results. To clean brushes, pour into a can and leave the brushes in it overnight. In the morning, wash out with warm water.

Rust and Ink Remover

Immerse the part of the fabric with the rust or ink spot alternately in solutions A and B, rinsing with water after each immersion.

A	Ammonium Sulfide Solution	1
	Water	19
B	* Oxalic Acid	1
	Water	19

* Poisonous.

Javelle Water
(Laundry Bleach)

Bleaching Powder	2	oz
Soda Ash	2	oz
Water	5	gal

Mix well until the reaction is completed. Allow to settle overnight and siphon off the clear liquid.

Liquid Laundry Blue

Prussian Blue	1
Distilled Water	32
* Oxalic Acid	1/4

Dissolve by mixing in a crock or wooden tub.

Glassine Paper

Paper is coated with or dipped in the following solution and then hung up to dry.

Copal Gum	10	oz
Alcohol	30	fl oz
Castor Oil	1	fl oz

Dissolve by letting stand overnight in a covered jar and stirring the next day.

Waterproofing Paper and Fiberboard

The following composition and method of application will make uncalendered paper, fiberboard, and similar porous material waterproof.

Paraffin (M.P. about 130 F)	22.5
Trihydroxyethylamine Stearate	3.0
Water	74.5

* Poisonous.

The paraffin wax is melted and the stearate added to it. The water is then heated to nearly the boiling point and vigorously agitated with a suitable mechanical stirring device while the mixture of melted wax and emulsifier is being slowly added. This mixture is cooled while it is stirred. The paper on fiberboard is coated on the side which is to be in contact with water. This method works most effectively on paper-pulp molded containers and has the advantage of being much cheaper than dipping in melted paraffin as only about one tenth as much paraffin is needed. In addition, the outside of the container is not greasy and can be printed on after treatment which is not the case when treating with melted wax.

*Waterproofing Liquid

Paraffin Wax	2/5 oz
Gum Dammar	1 1/5 oz
Pure Rubber	1/8 oz
Benzol	13 oz
Carbon Tetrachloride	To make 1 gal

Dissolve the rubber in the benzol, add the other ingredients, and allow to dissolve.

This liquid is suitable for wearing apparel and wood. It is applied by brushing on two or more coats, allowing each to dry before applying another coat. Apply outdoors as vapors are inflammable and toxic.

Waterproofing Heavy Canvas

Raw Linseed Oil	1	gal
Crude Beeswax	13	oz
White Lead	1	lb
Rosin	12	oz

Heat, while stirring, until all lumps are removed and apply warm to the upper side of the canvas, wetting it with a sponge on the underside before application.

---

* Inflammable.

Waterproofing Cement

China-Wood Oil	10 oz
Fatty Acids	10 oz
Paraffin Wax	10 oz
Kerosene	$2^1/_2$ gal

Stir until dissolved. Paint or spray on cement walls, which must be dry.

Oil- and Greaseproofing Paper and Fiberboard

This solution, applied by brush, spray, or dipping, will leave a thin film which is impervious to oil and grease. Applied to paper or fiber containers, it will enable them to retain oils and greases.

Starch	6.6
Caustic Soda	0.1
Glycerin	2.0
Sugar	0.6
Water	90.5
Sodium Salicylate	0.2

The caustic soda is dissolved in the water. Then the starch is made into a thick paste by adding a portion of this solution. The paste is then added to the water. The resulting mixture is placed on a water bath and heated to about 85 C, until all the starch granules have broken. The temperature is maintained about $^1/_2$ h longer at 85 C. The other substances are then added and thoroughly mixed. Less water may be used if applied hot and then a thicker coating will result.

Fireproof Paper

Ammonium Sulfate	8.00
Boric Acid	3.00
Borax	1.75
Water	100.00

The ingredients are mixed together in a gallon jug by shaking until dissolved.

The paper to be treated is dipped into this solution in a pan, until uniformly saturated. It is then taken out and hung up to dry. Wrinkles can be prevented by drying between cloths in a press.

Fireproofing Canvas

Ammonium Phosphate	1 lb
Ammonium Chloride	2 lb
Water	½ gal

Impregnate with the solution; squeeze out the excess, and dry. Washing or exposure to rain will remove fireproofing salts.

Fireproofing Light Fabrics

Borax	10 oz
Boric Acid	8 oz
Water	½ gal

Impregnate, squeeze, and dry. Fabrics so impregnated must be treated again after washing or exposure to rain as the fireproofing salts wash out easily.

Dry Fire Extinguisher

Ammonium Sulfate	15
Sodium Bicarbonate	9
Ammonium Phosphate	1
Red Ochre	2
"Silex"	23

Use powdered substances only. Mix well and pass through a fine sieve. Pack in tight containers to prevent lumping.

Fire-Extinguishing Liquid

Carbon Tetrachloride	95
Solvent Naphtha	5

The naphtha minimizes the development of toxic fumes when extinguishing fires.

Fire Kindler

Rosin or Pitch	10
Sawdust	10 or more

Melt, mix, and cast in forms.

Solidified Gasoline

* Gasoline	1/2 gal
Fine-Shaved White Soap	12 oz
Water	1 pt
Ammonia	5 oz

Heat the water, add the soap, mix and, when cool, add the ammonia. Then slowly work in the gasoline to form a semisolid mass.

Boiler Compound

Soda Ash	87
Trisodium Phosphate	10
Starch	1
Tannic Acid	2

Use powders, mix well, and then pass through a fine sieve.

Noncorrosive Soldering Flux

Powdered Rosin	1
Denatured Alcohol	4

Soak overnight and mix well.

* Inflammable.

Photographic Solutions

Developing Solution

Stock Solution A

Pyro	4 oz
Pure Sodium Bisulfite	280 gr
Potassium Bromide	32 gr
Distilled Water	64 oz

Dissolve in a glass or enamel dish.

Stock Solution B

Pure Sodium Sulfite	7 oz
Pure Sodium Carbonate	5 oz
Distilled Water	64 oz

Dissolve separately in a glass or enamel dish.
Use the following proportions:

Stock Solution A	2
Stock Solution B	2
Distilled Water	16

At 65 F, this developer requires about 8 min.

Acid-Hardening Fixing Bath

A	Sodium Hyposulfite	32
	Distilled Water	8

Stir until dissolved and then add the following chemicals in the order given, stirring each until dissolved:

B	Warm Distilled Water	2½
	Pure Sodium Sulfite	½
	Pure Acetic Acid (28%)	1½
	Potassium Alum Powder	½

Add B to A and store in dark bottles away from light.

Chapter II
ADHESIVES

Water-Dispersed Pressure-Sensitive Adhesive

Kraton® 1107		100
Xylene		75
Indopol® L-14		75
Amoco® Resin 18-210		40
Emulsifiers:		8
Triton® X-100	2	
Brij® 98	2	
Tween® 80	2	
Irganox® 1010		2
Water		90
Tackifier solution:		
Wingtack® 95		285
Hexane		95

Heat **Amoco® Resin 18-210** in solvent at 80 C (176 F) until dissolved. Add **Indopol® L-14** preheated to 80 C (176 F) and stir. Pour solution over **Kraton® 1107** and heat 15 min in an oven at 99 C (210 F). Mix and return to oven. Alternately heat and mix until homogeneous. Mix on a high-speed stirrer at 2000 rpm. Gradually add hot (99 C, 210 F) mixture of **Triton® X-100, Brij® 98** and **Tween® 80.** Add hot (99 C, 210 F) water slowly until inversion occurs. Add remaining water. Mix for 10 min. Blend in tackifier.

Physical Properties:

Rolling ball tack* — 6.0 cm
Quick stick adhesion* — 1.0 kN/m
Peel adhesion* — 1.5 kN/m

---

* Using Grade A airplane fabric as the substrate.

Emulsion stability—shelf life, days to separation at R.T.
100% undiluted — 60+
10% dilution — 10+
Freeze-thaw cycles* passed — 10

Hot-Melt Pressure-Sensitive Adhesive

Formula No. 1

Solprene 418	100
Wingtack® 95	150
Irganox® 1010	2
Amoco® Polybutene	0–125
Amoco® Resin 18-210	40

	No. 2	No. 3	No. 4	No. 5	No. 6
Kraton® 1107	100	100	100	100	
Wingtack® 95	100	100	100	100	100
Amoco® Resin 18-210	—	—	—	—	50
Amoco Resin 18-290	40	40	20	40	—
Amoco Polybutene L-14	60	—	—	—	—
Amoco Polybutene H-100	—	10	—	—	—
Amoco Polybutene H-1500	—	—	40	40	20
Irganox® 1010	1	1	1	1	—
Antioxidant 330	—	—	—	—	5
Physical Properties:					
Rolling ball tack, PSTC-6 (cm)	2.3	—	—	—	19.3
Quick stick adhesion, PSTC-5 (lb)	2.0	1.8	5.3	4.5	—
Peel adhesion, PSTC-1 (lb)	2.5	5.8	10.7	12.0	2.4
Shear adhesion, Amoco test (s)	1.6	263.0	37.6	32.3	3309

* 8 h @ –12 C (10 F), 16 h @ R.T.

No. 7

Kraton® 1107	100.0
Wingtack® 95	100.0
Irganox® 1010	1.5
Shellflex® Oil 371-N	10–80
Cumar® LX-509	20–80

Hot-Melt Glazing Compound

Thermoplastic Elastomer	3.84
Tackifying Resin	7.69
Amoco® Resin 18-290	9.61
Amoco® Polybutene H-1500	24.98
Calcium Carbonate	42.27
Silica Extender	9.61
Titanium Dioxide	1.92
Antioxidant	0.04
Stabilizer	0.04

In the laboratory the compound was prepared in a sigma blade mixer equipped with a jacket for steam and cold water using the following steps:

Start	Steam on to 149 C (300 F); charge polybutene, antioxidant, stabilizer, calcium
5 min	Add copolymer
10 min	Add Resin 18
30 min	Add tackifying resin
65 min	Dump

Note: This nonskinning sealant is formulated for application by melting with a portable hand-extruder. In the molten form it flows easily and quickly into channels. This hot extrusion dries the joint, makes year-round application possible, and improves adhesion.

Physical Properties:
AAMA 804.1 — Requirements
Hardness, Shore A — ≤ 55
Normal — 40
21 days at 60 C (140 F) — 41
21 days UV — 46
Peel Adhesion, R.T. — ≥ 0.70 N/mm
Cohesive Failure — 90%

Tape Sealant

Formula No. 1

(Partially Crosslinked, Butyl)

Butyl Elastomer	34.25
Amoco® Polybutene H-100	34.25
Stearic Acid	0.68
Carbon Black Beads	30.82

The sealant is mixed in a sigma blade mixer equipped with compression ram and a jacket for steam and cold water using the following steps:

Start	Cold water on; charge rubber, stearic acid, $^1/_2$ carbon black, $^1/_2$ polybutene; mix
5 min	Add $^1/_4$ carbon black, $^1/_4$ polybutene
10 min	Add $^1/_4$ carbon black, $^1/_8$ polybutene
20 min	Cold water off; steam to 120 C (250 F)
30 min	Steam off; cold water on
35 min	Add remaining polybutene
40 min	Dump.

Note: High performance, application ease and low cost are among the advantages of windshield and backlight sealant tapes based on partially crosslinked butyl elastomers. This formulation combines properties which normally tend to be mutually exclusive: softness and resistance to flow, resistance to heat aging and low temperature flexibility, outstanding adhesion and cohesion without staining.

	No. 2	No. 3
	(Polyisobutylene)	
Amoco® Polybutene H-300	19.90	18.90
Polyisobutylene Rubber	19.90	18.90
Butyl Rubber	—	5.25
Calcium Carbonate	35.05	35.05
Platelet Talc	7.50	6.50
Attapulgas Clay	12.25	11.00
Diatomaceous Silica	3.40	2.40
Titanium Dioxide	2.00	2.00

In the laboratory the polyisobutylene and butyl rubbers are mixed in a dual-arm sigma blade mixer for approximately 20 min. Dry ingredients are incrementally added in the order listed while mixing. The sealant is dumped, rested 24 h, and extruded.

Note: Formula No. 2 is a polyisobutylene-based tape which offers the high performance of conventional asbestos-containing sealants at a lower raw material cost while eliminating the hazards associated with asbestos. Formula No. 3 incorporates butyl rubber. These sealants are suggested for use in metal building systems and automotive applications.

No. 4

(EPDM/Polybutene)

Ethylene Propylene Terpolymer	40.40
Amoco® Polybutene H-300	40.40
Amoco® Resin 18-210	8.08
Zinc Oxide	2.02
Stearic acid	0.41
Tetraethylthiuram Disulfide	0.27
2-Mercaptobenzothiazole	0.13
Sulfur	0.21
Carbon Black (FEF)	8.08

The EPDM is banded on a tight mill at 82 C (180 F) to 93 C (200 F) for 10 min. An **H-300** Polybutene/carbon black slurry is prepared, added in increments and milled 30 min. Zinc oxide and stearic acid are added and dispersed (5–10 min). **Resin 18-210** is added and dispersed (5–10 min). Accelerators are added and dispersed (5 min). Sulfur is added and milled for 5 min. Stock is then extruded into 1/2 in. × 1/2 in. tape and cured in a forced draft oven for 24 h before testing.

Note: Outstanding ozone and weather resistance suggest this formulation for automotive windshield and "high-rise" curtain wall applications. The compound uses a partially vulcanized ethylene propylene terpolymer extended and plasticized with **H-300** Polybutene.

No. 5

(Polybutene, Nondrying)

Amoco® H-300 Polybutene	24.01
Amorphous Polypropylene Homopolymer	5.05
Butyl Rubber	1.72
Clay	16.37
Calcium Carbonate	44.05
Diatomaceous Silica	4.00
Cotton Fiber	4.80

In the laboratory, the sealant is compounded in a sigma blade mixer. Amorphous polypropylene is premixed with twice its weight of polybutene to facilitate complete dispersion. (This premixing may not be required in commercial scale equipment.) Additions to the mixer are made in the order of listing of materials. The entire mass is then mixed for 1 h after the last addition.

Note: **Amoco H-300 Polybutene** contributes oxidation resistance and permanent pliability to this nondrying sealant tape formulated without asbestos. This type of sealant should find application in concealed joints for mobile homes, recreational vehicles, and trailers.

No. 6

(Butyl)

Amoco® H-300 Polybutene	32.0
Butyl Rubber	8.0
Calcium Carbonate	35.0
Attapulgas Clay	16.0
Diatomaceous Silica	5.0
Titanium Dioxide	4.0

In the laboratory the butyl rubber and polybutene are mixed in a dual-arm sigma blade mixer for approximately 20 min. Dry ingredients are incrementally added in the order listed while mixing. The sealant is dumped, rested 24 h, and extruded.

Note: This sealant is a low-cost alternative to asbestos-containing systems and is suggested for use in metal building systems and automotive applications.

Insulated Glass Sealant

	No. 1	No. 2
	(Butyl)	
Butyl Elastomer	15.6	16.6
Carbon Black (HAF N330)	23.5	25.5
Phenolic Resin	29.5	25.6
Amoco® Polybutene H-1900	23.5	17.5
Amoco® Resin 18-290	—	5.9
Ethylene Propylene Rubber	7.9	8.9

Note: Formula No. 1 is a hot-melt applied sealant demonstrating excellent properties in glazing applications where the antifogging properties yielded by **H-1900** Polybutene are desirable. Formula No. 2 adds **Amoco Resin 18** producing a sealant with increased lap shear and dead load properties.

Physical Properties:

Lap Shear* (kPa)	407	427
Hardness, Shore A	54	64
Penetration, ASTM C-782 (mm)	12	7
Dead Load, 70 C, 500 g load (h to failure)	4.2	8+

*C = cohesive failure, A = adhesive failure

Shoe Sole Adhesive

	Formula No. 1	No. 2	No. 3
		(Polychloroprene)	
Rubber Mill Mix:			
Polychloroprene	100 kg	100 kg	100 kg
Rubber Antioxidant	2 kg	2 kg	2 kg
Magnesium Oxide	4 kg	4 kg	4 kg
Zinc Oxide	5 kg	5 kg	5 kg
Prereact:			
Phenolic Resin (heat reactive)	45 kg	45 kg	45 kg
Hexane 50%, Cyclohexane 50% (by volume)	403.8 l	403.8 l	403.8 l
Magnesium Oxide	4 kg	4 kg	4 kg
Water	1 kg	1 kg	1 kg
Dissolve:			
Prereacted Ingredients	above	above	above
Rubber Mill Mix (above)	111 kg	111 kg	111 kg
MEK 50%, Acetone 12.5%, Ethyl Acetate 37.5% (by volume)	269.2 l	269.2 l	269.2 l
Amoco® Resin 18-290	30 kg	40 kg	45 kg

The rubber mill mix is milled on a two-roll tile mill, banded and chopped into small pieces. The heat-reacted phenolic resin is prereacted in the hexane, cyclohexane, magnesium oxide and water for 24 h under agitation. Combine the mill mix and the prereacted phenolic resin with the remaining ingredients. Agitate the batch until completely dissolved in the solvent. Adhesive is now ready to be applied to the upper material and TPE shoe sole stock.

Airplane Wing Sealant

Formula No. 1

(Filleting)

Triazine Ring-Crosslinked 1,2,4-Oxadiazole Elastomer	1.00
Asbestos	0.05
Carbon Black	0.80
Epoxy Silane	0.02

No. 2

(Channel)

Triazine Ring-Crosslinked 1,2,4-Oxadiazole Elastomer	1.00
Asbestos	0.15
Metal Deactivator	0.05
Lithium Aluminum Silicate	0.55

Caulking and Sealant Compound

(Two-Component Elastomeric)

A	TDI	91.0
	3000 MW Triol	150.0
	Teracol 2900	429.0
	Calcined Clay	400.0
	Viplex 525	125.0
B	2000 MW Diol	600.0
	Cosan 31	0.5
	Calcined Clay	600.0
	Viplex 525	200.0

Gun-Grade Caulk

(Polybutene)

Amoco® H-100 Polybutene	20.00
Vinyltoluene/Vegetable Drying Oil Copolymer	4.39

Tall Oil Fatty Acid	0.49
Calcium Carbonate	50.80
Platelet Talc	5.80
Diatomaceous Silica	3.20
Hard Clay	11.70
Mineral Spirits	3.33
Cobalt Drier (6%)	0.29

In the laboratory the vehicle ingredients are blended in a dual-arm sigma blade mixer. Fillers are added incrementally in the order listed and then mixed 30 min. The mixture is thinned to desired viscosity with mineral spirits, the driers added, and the caulk mixed to a smooth consistency.

Note: This gun-grade caulk uses a vehicle containing polybutenes and a small amount of vinyltoluene/vegetable drying oil copolymer. A combination of fillers is then added to produce a caulk that offers substantial cost savings and meets federal specifications. An additional advantage of this caulk is that it utilizes no potentially hazardous fibrous materials.

Acrylic Latex Roof Mastic

Natrosol® 250 MXR (3)%	140.74
Water	16.0
Tamol® 850	4.76
Potassium Tripolyphosphate	1.43
Ethylene Glycol	24.38
Nopco® NXZ	1.90
Ti-Pure® R-960	70.37
Duramite®	422.72
Kadox® 515	46.95

Grind on high-speed disperser for 15 min. Then let down with the following ingredients:

Rhoplex® EC-1791	470.59
Texanol®	6.95
Skane® M-8	2.10
Nopco® NXZ	1.90
Ammonium Hydroxide (28% ammonia)	0.95

Grind for an additional 10 min.

Physical Properties:

P/B Ratio — 2.08/1.0
Wt./Gal— 12.1 lb
PVC — 43.0%
NV Vol. — 51.4%
Total Solids — 66.9%

Note: Should be applied in two or more coats to a total dry film thickness of 20–30 mils with proper drying between coats. This application recommendation is designed to help avoid the possibility of pinholing and mudcracking which can be associated with extremely thick deposits of wet coatings.

Chapter III

BEVERAGES AND FOODS

Spray-Dried Orange Juice Drink Mix

Orange Juice Concentrate	30.00
Maltrin® M100 Maltodextrin	28.50
Water	40.00
Citric Acid	1.00
Calcium Phosphate	0.12
Sodium Citrate	0.12
Orange Flavor	0.20
Aspartame	0.06

Mix all ingredients thoroughly. Pasteurize by high temperature short time method. Dry using as low an outlet temperature as possible.

This formulation, using orange juice concentrate (42% solids) is made up at 40% total solids. It is dried on a Niro utility dryer to a 3% moisture level then cooled and packaged. Using a dryer with cold jacket chamber walls may allow the use of a higher percentage of orange juice.

Usage: **Maltrin® M100** maltodextrin is an excellent carrier for hard-to-dry fruit juice such as orange juice.

Benefits of **Maltrin® M100** maltodextrin: (1) Low hygroscopicity, (2) High solubility, (3) No flavor addition from carrier.

Dry-Mix Chocolate Beverage

Lecithinized Cocoa	19.7
Sugar*	78.8
Lactarin MV 306	1.5

* When replacing sugar with artificial sweeteners, the mix may require instantizing to minimize lumping when added to milk.

Add 17.77 g of dry mix blend to 250 ml of cold milk. Spoon stir for 1 min.

Chocolate Milk

Formula No. 1

Milk (2% BF)	93.28
Sugar	6.00
Cocoa	0.70–1.30
SeaKem® CM 611	0.02–0.03

Notes: Shelf life of two months—0.02%–0.025% **CM 611** and process at 143–149 C (290–300 F) for 3–4 s.

Shelf life of six months—0.02%–0.025% **CM 611** and process at 149 C (300 F) for 7 s.

144 C for 16 s recommended to destroy enzymes responsible for "sweet curdling."

No. 2

Partially Skim Milk	92.38
Sugar	6.00
Cocoa	1.60
Carrageenan	0.02

No. 3

(Filled)

Water	81.45
Sucrose	6.00
Skim Milk Powder	9.00
Vegetable Fat	2.00
Cocoa	1.20
Emulsifier	0.30
Carrageenan	0.30
Flavor	as required

No. 4

(Pasteurized)

Milk (2%)	89.895
Sucrose	5.930
Water	3.350
Dutch Cocoa	0.800
SeaKem® Carrageenan	0.025–0.040

Dry blend cocoa, sucrose, and carrageenan and add to milk. If liquid sugar or corn syrup is used, meter into batch. Add dry blended ingredients. Pasteurize by batch. Cool to 40 F (40 C) and fill.

Chocolate Syrup

	Formula No. 1 *(66% Solids)*	No. 2 *(45% Solids)*
Sucrose	38.90	28.60
Corn Syrup (high fructose)	26.40	26.40
Cocoa	9.00	9.00
Lactarin MV 308	0.32	0.50
Benzoic Acid	0.10	0.10
Potassium Sorbate	0.10	0.10
Vanilla	0.05	0.05

Dry blend 1 part **Lactarin** carrageenan with 10 parts sugar. Disperse into 75% of water in formula. Heat to 85–90 C. Combine high fructose corn syrup, remaining sugar and cocoa in separate container. Heat to 85–90 C. Combine the contents of the two containers by pouring or pumping carrageenan dispersion into the cocoa system. Maintain at 85–90 C for 25 min. Add the benzoic acid and potassium sorbate to the remaining water. Heat to 85 C before adding to cocoa system. Mix and hold entire system for 5 min. Retare to original weight before filling. Lower total solids system should be hot filled. High total solids system may be cooled to 55 C before filling.

Egg Nog

Formula No. 1

Condensed Skim Milk	37.00
Fluid Milk (4% fat)	34.84
Heavy Cream (40% fat)	13.80
Corn Syrup No. 55 (high fructose)	8.30
Flavor and Spice	5.00
Egg Yolk Solids	1.00
SeaKem® CM 611*	0.10

Note: Use Standard HTST or batch process.

No. 2

Milkfat	6.00
Milk Solids (nonfat)	12.00
Sucrose	12.00
Egg Yolk Solids	0.50
Emulsifier	0.20
Carrageenan	0.10
Flavor and Color	as required

Papaya Drink

Water	6 cups
Papaya Puree	3 cups
Sugar	3/4 cup
Citric Acid (substitute lemon or lime juice to taste)	1/2 tsp.

Mix ingredients together and serve cold.

* If xantham gum is used to provide additional body the use level should be 0.05–0.07.

Soft Drink

Demineralized Water	90.25
Sucrose	7.30
Fruit Concentrate	2.00
Citric Acid	0.15
Ascorbic Acid	0.10
Alginate	0.12
Flavor	as required

Margarita Mix

Sucrose	74.4
Powdered Lime Juice	6.0
Powdered Lemon Juice	6.0
Powdered Orange Juice	6.0
Lime Oil—Encapsulated	0.5
Lemon Oil—Encapsulated	0.5
Orange Oil—Encapsulated	0.5
Citric Acid	4.0
Sodium Carboxymethyl Cellulose	0.4
Sodium Citrate	1.5
Tricalcium Phosphate	0.1
Instant Dried Egg White **Hentex P-600**	0.1
BHA	trace

Imitation Cheese Block

Water	50.73
Shortening	25.00
Sodium Caseinate	11.00
Maltodextrin (10 DE)	5.32
$CaSO_4\ 2H_2O$	3.00
Gelcarin® GP 911	2.50
Sodium Chloride	2.00
Adipic Acid	0.25
Potassium Sorbate	0.20
Color and Flavor	to suit

Melt shortening in a steam jacketed kettle with a surface scraping stirrer such as a Stephan Cooker by raising its temperature to 85–90 C. This type of kettle allows the introduction of all ingredients at once without concern of lumping or solubility problems.

Dry blend the dry ingredients homogeneously and gradually add this mix into the hot shortening with continuous and adequate agitation. Considerable lumping may occur if this dry-blend is added too fast. Make sure ingredients are completely dissolved.

Heat the water up 85–90 C, then slowly add into the shortening/mix while constantly agitating. Maintain temperature at 85 C.

After a smooth and homogeneous mix is obtained, pour contents into containers and store.

Note: Flavor may be added just prior to filling to minimize its loss during heating.

Cheese Dip Mild Cheddar Cheese

Insta Starch 5/70	18.8
Centennial II	31.4
Cheddease 180 Cheese Powder	14.0
Cheddease 270 Cheese Powder	22.0
Romano L. Cheese Powder	2.0
Flour Salt	2.0
Disodium Phospahte	0.2
Tetrasodium Phosphate	0.7
Natural Cheese Flavor 4891 122 5	3.8
Lactic Acid Powder 60%	1.4
Monosodium Glutamate	0.2
Insta Thick® X-10	3.2
Mono and Diglycerides **G695**	0.5

Add the mono and diglycerides to the flour salt and blend. Add 30% of the **Insta Starch** to the flour salt and mono-diglyceride mixture and blend thoroughly. Add remaining ingredients and mix thoroughly using a tumble type blender.

Preparation: Add 88.2 g to one cup of cold whole milk and mix for 2 min using a wire whip. Refrigerate for 5 min before serving. (For a smoother appearance, mix again for 1 min after refrigeration.) The dip can be heated in the microwave.

Cheese Spread/Sauce

Cheddar Cheese	40.0
RT-50-S (50% protein)	8.4
Sodium Chloride	1.0
Skim Milk (1/2% fat)	49.9
Annatto	to suit
Lactarin XP 3074	0.6

Dry blend **Lactarin XP 3074** carrageenan, salt and 2% of the 8.4% whey protein. Add remaining whey protein to skim milk in jacketed kettle with agitation. When using an open kettle, add 5% additional water for evaporation. Add dry ingredients with agitation. Heat while agitating to 85 C. In a separate vessel, with agitation, melt cheddar cheese to 85 C. Add melted cheese to contents of first kettle while maintaining agitation and 85 C temperature. Add color and flavor. Mix until uniform. Fill and cool.

Note: A Stephan Cooker, which eliminates the need for a two-kettle system, is desirable. Otherwide, a two-vessel system is necessary.

Sour Cream and Dip

Locust Bean Gum	41
Dextrose	34
Gelcarin® GP 359	25

Standardize fresh cream to 18–20% milk fat. Blend stabilizer into cream. Homogenize, pasteurize and cool to 72 F. Inoculate with lactic acid starter culture. Incubate 14–16 h to a titratable acidity of 0.7% or a pH of 4.5. Cool to 40 F. Package.

Note: The respective ratio of carrageenan to locust bean gum can vary between 2:1 and 1:2 depending on desired body of the finished product. Typical use levels of the sour cream stabilizer blend in the finished product should range from 0.1–0.2%.

This standard procedure can be appropriately modified for hot pack or directly acidified sour cream production.

Dry Mix Dressing

Formula No. 1

(Blue Cheese Flavored)

Maltodextrin	31.0
Micron Natural Blue Cheese Flavor **TAK-50406**	30.0
Micron Natural Cream Cheese Flavor **TAK-50560**	15.0
Instant **Cleargel**	15.0
Salt (fine flake)	6.0
Citric Acid	2.0
White Pepper (ground)	0.4
Micron Natural Lemon (5X) **TAK-40606**	0.2
Micron Natural Onion **TAK-50363**	0.2
Blue Cheese Color Blend:	0.2
Maltodextrin	99.0
Blue #1	0.7
Yellow #6	0.3

Combine all ingredients and mix well. Use 50.0 g blue cheese salad dressing mix to make one pint (two cups) of salad dressing.

Preparation: Mix together one cup (8 oz) mayonnaise, one cup (8 oz) buttermilk, and one envelope (50 g) blue cheese salad dressing mix. Stir well.

No. 2

(Italian)

	(Household)	*(Metric)*
Vegetable Oil	1/2 cup	182.5 g
White Vinegar (5% acidity)	1/4 cup	100.0 g
Water	1/4 cup	100.0 g
Viscarin® SD 339 Carrageenan	0.1 oz	2.0 g
Seasoning Mix #9849-A-36	0.1 oz or to taste	2.0 g or to taste

Combine liquid ingredients by stirring until thoroughly mixed. Slowly add the blended spices and **Viscarin** to the liquid phase. Continue mixing until the liquid has thickened. After use, refrigerate the remaining salad dressing for better stability.

Note: The resultant salad dressing is at a pH range of 3.4–4.0 which provides tartness and mouthfeel typical of oil and vinegar systems. Increased amounts of oil or spices may require a higher level of **Viscarin** carrageenan.

Low Calorie Pourable Salad Dressing

(Creamy Italian)

A	**Maltrin® M040** Maltodextrin	16.30
	Salt	3.00
	Buttermilk Solids	2.50
	Merecol FA	1.25
	Onion Powder	0.70
	Garlic Powder	0.60
	White Pepper	0.20
	Sodium Benzoate	0.10
	Potassium Sorbate	0.10
	Aspartame	0.02
B	Vinegar (5% acidity)	15.00
	Lemon Juice	0.50
	Water	51.93
C	Soybean Oil	7.50
	Polysorbate 60	0.30

Blend Part A dry ingredients. Add Part A ingredients slowly to water, using high shear agitation, until uniformly dispersed, approximately 2–3 min.

Next add vinegar and lemon juice to the above mixture and incorporate thoroughly. Add oil and polysorbate 60 (Part C ingredients). Incorporate thoroughly, about 5 min mixing. Mixing speed may need to be increased. Heat mixture to pasteurization. Homogenize at 1000 psi, single-stage homogenizer. Package aseptically.

Note: **Maltrin® M040** allows the oil level to be lowered very significantly. This formulation has 1.5 calories/gram. This formulation has one-half the calories and one-quarter of the oil level of a standard salad dressing.

Low-Oil, Pourable, Italian-Type Dressing

Water	66.4
White Vinegar (5% HOAc)	24.0
Sucrose	6.0
Salt	3.2
Viscarin® SD 389 Carrageenan	0.2
Xanthan Gum	0.2
Herbs and Spices	0.2
Optional—Vegetable Oil	(up to 2.0)

Cold Process:

Part A: Combine vinegar with 75% of available water in a mixing tank. Dry blend the stabilizer and sugar before adding to vinegar/water solution. Agitate until stabilizers are completely hydrated.

Part B: In separate tanks, dissolve salt and hydrate herb and spice in remaining (25%) water.

Combine Parts A and B while rapidly agitating until uniform thickening is achieved. Avoid air incorporation. Fill into containers.

Notes: Acetic acid may be used instead of vinegar. The degree of substitution will depend upon desired acidity level. The volume difference can then be adjusted by additional water.

At the pH range of 2.7–3.0, there is sufficient tartness which is typical of salad dressings. Sweetness is contributed by the sugar.

Pigments from the herbs and spices can be removed by prehydration to avoid color migration and cloud.

This formulation and procedure are general suggestions. The ratios of **Viscarin** and xanthan gum can be adjusted to provide different textural and suspension characteristics.

Up to 2% vegetable oil may be used for organoleptic reasons. While **Viscarin** is not an emulsifier, it will suspend the oil-in-water mixture indefinitely and prevent its separation.

Cottage Cheese Dressing

	Formula No. 1 *(High Fat 17.4)*	No. 2 *(Medium Fat 5.7)*	No. 3 *(Low Fat 0.67)*
Whole Milk (3.7% fat)	54.5	90.1	6.1
Heavy Sweet Cream (40% fat)	38.4	6.0	—
Skim Milk (0.5% fat)	—	—	90.0
Skim Milk Powder	3.2	—	—
Sodium Chloride	3.8	3.8	3.8
Gelcarin® GP 911	0.1–0.5	0.1–0.5	0.1–0.5

Disperse milk solids in fluid milk. Blend carrageenan and salt. Add to milk using good agitation. Add heavy cream. Pasteurize dressing mix at 160 F and cool to 40 F. Creaming mixture should be made at least 24 h before application to curds. Use dressing at rate of one part dressing to three parts unsalted curd (25% dressing: 75% curd). Pump dressing over cold curds in cheese vat and agitate slowly until dressing is incorporated uniformly around the curd. Fill containers with finished creamed cottage cheese.

Note: Alternate carrageenan (**SeaKem® IC 611**) will provide additional shear stability and can be used in higher levels to provide body and cling without gelation.

In practice milk solids, fluid milk, and cream may be mixed in various portions to obtain desired fat and solids level. Citric acid may be used at approximately 0.10% to adjust the pH of the dressing to about 5.5.

Béchamel Sauce

Milk	88.00
Butterfat	8.00
Modified Starch	2.50
Emulsifier	0.40
Carrageenan or Carrageenan/Guar Gum	0.40
Salt and Spices	as required

Barbecue Sauce

Formula No. 1

(Dry Mix)

A	Tomato Powder	38.0
	Brown Sugar	37.0
	Spice Mix (or to suit)	16.0
	Pure-Dent B700 Corn Starch	9.0
B	Water	65.7
	Vinegar	9.1
	Mix (from Part A)	25.2

Dry blend ingredients in Part A. Add water to mix, heat with stirring to boil. Cool to 140 F. Add vinegar and blend.

No. 2

Tomato Catsup	201.0	g
Worcestershire Sauce	14.0	g
Chili Powder	1.7	g
Sugar	3.0	g
Orange Juice Concentrate	22.0	g
Tabasco Sauce	0.2	g
Red Miso	150.0	g
Water	225.0	ml

All ingredients are first mixed in a Waring Blender. The mixture is brought to a boil and then simmered for 15–20 min. The sauce obtained has the consistency and color of conventional barbecue sauce, has 32.6% of reflectance, 16.6 of red values, and 17.7 of yellow values, as measured by a Hunterlab Model D25, color and color-difference meter. It contains 2.9% protein and 77% moisture.

Note: If red miso is substituted for some of the tomato paste and the salt is omitted, the modified product is delicious on hamburger and other meats or as a barbecue sauce in which to cook and base chicken, pork, or beef. The ability to substitute other ingredients in this sauce is almost unlimited.

No. 3

Tomato Paste (30% solids)	27.50
Water	21.75
Cider Vinegar (50 grain)	22.00
Granulated Brown Sugar	8.20
Granulated White Sugar	8.20
Soy Sauce	5.00
Hickory Smoke Flavoring #333	2.00
Granulated Salt	2.00
Methocel® A4M Premium	1.00
Thermo-Flo Modified Food Starch	1.00
Chili Powder	0.55
Mustard Flour	0.20
Red Pepper (ground, 20,000 H.U.)	0.20
Toasted Onion Powder	0.15
Garlic Powder	0.15
Cinnamon (ground Korintje)	0.10

Prepare a starch slurry with one-half of the water; cook. Blend **Methocel** with dry ingredients other than sugars. Mix all ingredients except sugars with the starch solution. Cool to hydrate **Methocel** fully. Add sugars last.

Shrimp Cocktail Sauce

(30% Solids)

Chili Sauce	11.7
Catsup	35.2
Tomato Paste	23.4
Mr. and Mrs. "T" Bloody Mary Mix	11.7
Horseradish	2.3
Lemon Juice	5.9
Celery Salt	0.5
Onion Powder	0.5
Garlic Salt	0.5
Red Pepper (20,000 H.U.)	.4
Sugar	7.0
Methocel® A4M Premium	0.9

Combine chili sauce, catsup, Bloody Mary Mix, horseradish, lemon juice, and tomato paste and mix thoroughly with moderate agitation. Thoroughly dry blend spices, sugar, and **Methocel**, then add dry blend to liquid phase using moderate agitation. Continue to mix for 5 min. Cool and store in refrigerator.

Au Jus Gravy Mix

Maltrin® M100 Maltodextrin	25.20
Corn Starch	17.00
HVP **Type BE Powder**	14.00
Monosodium Glutamate	12.00
Salt (flour fine)	8.00
Sucrose	6.00
Autolyzed Yeast Extract Powder (standard dark)	6.00
Hydrogenated Vegetable Oil or Beef Fat	5.00
Caramel Color Powder	4.00
Onion Powder	2.00
Celery Seed (micro-milled)	0.30
Paprika 50-A	0.30
Ground Black Pepper	0.20

Thoroughly blend dry ingredients. Mix 20 g of dry mix with1 cup of water in saucepan. Bring to boil while stirring; simmer for 1 min on low heat.

Chicken Soup Mix

(Dry Mix)

A	**Maltrin® M100** Maltodextrin	36.11
	Salt (flour fine)	30.00
	Monosodium Glutamate	10.00
	Chicken Fat	8.00
	Granulated Sugar (extra-fine)	6.00
	Chicken Meat (dried)	3.00
	Chicken Soup Spice Mix	3.00
	Onion Powder	3.00
	Soluble Pepper	0.60
	FD&C Color Yellow No. 5	0.01

B	Parsley Flakes	0.28

Blend Part A thoroughly. Then run through a mill to get uniform particle size. Add Part B and mix gently.

Preparation of Soup: To 1 qt of boiling water add 28.35 g of dry mix. Let simmer for 5 min.

Walnut Crackers

Cleargel Starch	4.286
Baka-Snack	4.286
Special Corn Flour	14.286
California Walnut Meal	22.857
Fine Walnut Pieces	5.714
Cinnamon	0.857
Salt	1.714
Sugar (fine granulated)	28.571
Water	17.143
Yeast	0.286

Mix cinnamon, salt, and sugar for 1 min. Add walnuts, starches, and special flour and mix for 2 min. Dissolve yeast in water. Add water mix until pickup. Ferment for 2 h. Sheet $^1/_8$-in. thick. Bake at 325 F for 5 min. Reduce heat to 250 F. Bake 14–15 min.

Raisin Cracker

A	Whole Wheat Flour	26.96
	Cake Flour	15.05
	Baking Powder	0.43
	Baking Soda	0.27
B	All-Purpose Shortening	10.21
	Raisin Paste	18.70
	Brown Sugar	4.84
	Honey	11.93
C	Vanilla (real)	0.43
	Whole Milk	11.18

Premix Part A and add to premixed sugar and shortening from Part B. Mix Part B ingredients together and cream well. Add Part C ingredients last. Mix for 8 min at a slow speed. Allow dough to ferment in bowl for 2 h at 75–80 F. Sheet to $^1/_4$-in. thick. Cut crackers approximately 1 in. × 3 in. Bake at 375 F for 5 min. Reduce to 275 F and bake for 10 min.

Notes: Coarse sugar and egg wash can be applied to top before baking. Sesame seeds can be used in formula and on surface. Dough can be chilled overnight and held for later use. More raisins may be added to dough by using 10% chopped raisins. These are not to exceed level of raisin paste.

Papaya Bread

Papaya Puree	3	cups
Eggs	3	
Shortening	$^3/_4$	cup
Flour (all purpose)	4	cups
White Granulated Sugar (C&H)	1	cup
Baking Soda	1	T
Salt	1	tsp
Cinnamon	1	tsp
Nutmeg	$^1/_2$	tsp
Chopped Walnuts	1	cup

Preheat oven to 325 F. Spray bridgefold glazed mini pans with Pam (aerosol shortening spray); set aside. Place papaya puree, shortening, and eggs into mixing bowl. Mix on slow speed to break eggs and combine liquids, approximately 30 s. Add flour, sugar, baking soda, salt, nutmeg, and cinnamon to liquid mix. Beat $1^1/_2$ to 2 min on medium speed until all ingredients are thoroughly mixed. Use no. 8 scoop to scale papya bread mix into mini loaf pan, 1 scoop per section. Bake 55–60 min. Cool on rack. Yield: 12 4 oz loaves.

"Your Basic" Walnut Bread

Patent Flour	16	lb
Water	1	gal
Salt	4	oz
Low-fat Dry Milk	8	oz

Sugar	1	lb
Shortening	1	lb
Eggs	1	lb
California Walnuts (diced or bits)	8	lb

Mix all ingredients until incorporated, adding walnuts in the final stage to ensure even distribution. Allow 1 h fermentation. Scale at desired loaf weights: suggested range is 1 lb 4 oz (20 oz). Rest dough 15 min. Make up and proof. Bake at 375 F (109 C).

Tempura Batter

Micro-Crisp	50.0
Micro-Crisp D	10.0
Wheat Flour (all-purpose)	37.0
Sodium Aluminum Phosphate	1.5
Sodium Bicarbonate	1.5

Preparation for Chicken: Add 160 ml water to 100 g dry mix. Parboil fresh chicken for approximately 15–20 min. Rinse chicken in cool water. Predust with **Textaid-A.** Dip in batter. Parfry 60 s at 385–400 F. Freeze immediately. Microwave frozen pieces. Internal temperature of the meat should be 150 ± 10 F after microwaving.

Preparation for Fish: Add 180 ml water to 100 g dry mix. Dip frozen fish into warm water. Coat fish with **Textaid-A.** Dip in batter. Coat with Japanese bread crumbs. Parfry 60 s at 380–400 F. Freeze immediately. Microwave. Internal temperature of fish should be 150 F after microwaving.

Note: $3\,^{3}/_{4} \times 3\,^{1}/_{8} \times ^{5}/_{16}$ in., 2 oz frozen cod fillets are used.

Whipped Topping

Formula No. 1

(Ready-to-Use)

Methocel® K100-LV Premium	0.5
Xanthan Gum	0.1
Water	58.7
Dariplus C	25.0

Tween® 60	0.4
Drewpol 6-2-S	0.1
Flavoring (sugar, etc.)	15.0
Salt	0.1

Note: Other low viscosity **Methocel** products such as **Methocel F50-LV Premium** or **Methocel K35 Premium** are also used in this type of recipe. Other suitable emulsifiers may also be used in this recipe.

No. 2

(Concentrated)

Paramount C or **Dariplus C**	38.00
Durfax 60	0.55
Durtan 60	0.20
Methocel® F50-LV Premium	0.23
Methocel® K15M Premium	0.09
Kelgin LV	0.08
Sugar	22.00
Flavor and Color	as desired
Water	38.85

Preparation Method 1: For ease of handling and for complete stabilizer hydration, the **Methocel** gum and xanthan powders should be dispersed in **hot** water (160–190 F) $^1/_5$ to $^2/_5$ of the total water and agitated thoroughly to "wet out" and separate the individual particles. The remainder of the water should be added as **cold** water and agitation continued until all of the particles of **Methocel** are completely hydrated. This ordinarily requires 10–30 min.

The fat and other emulsifiers should be comelted at minimum temperatures and added with sufficient agitation to disperse well into the water. It may be necessary to raise the temperature of the water to prevent "shock out" of the oil. Continue mixing until a uniform mix is obtained. The sweeteners, salt, and flavoring are then added and agitated until completely dissolved.

This mix may then be pasteurized at 165 F for 30 min under mild agitation. The pasteurized mix can be homogenized into a two-stage homogenizer at 2000 psi in the first stage and at 500 psi in the second stage. Homogenization should be followed by rapid cooling to prevent possible reagglomeration of the fat particles. If the product is to be frozen, it should be placed in the freezer immediately. Frozen emulsions should be placed in a 40 F refrigerator for thawing (usually 6 h) before whipping.

Preparation Method 2: Dry blend the **Methocel** gum and xanthan gum and then disperse in formulation water under agitation. Agitate until completely solubilized. Raise temperature of the gum solution to approximately 120 F and add sugar under agitation, continuing until solubilized. Co-melt fat and emulsifiers and add under agitation to the water phase. Continue mixing until a uniform mixture is obtained. Proceed with addition of remaining flavorings, homogenization, and quick chilling as discussed in Method 1.

No. 3

(Long Life Liquid—30% Fat Content)

Cream (35% fat content)	87.50
Liquid Buttermilk	12.00
Skim Milk Powder	0.50
Carrageenan	0.20

Vienna Cream Dessert with Whipped Cream

Milk (2% fat)	82.75
Sugar	12.00
Cocoa	3.00
Corn Starch*	2.00
SeaGel® DP 437*	0.25

Cold fill process.

Flan with Syrup

Milk (2% fat)	83.30
Sugar	12.00
Cocoa	3.00
Corn Starch*	1.50
SeaGel® FL 674P*	0.20

* Carrageenan starch rations must be balanced to adjust texture and other desired properties.

Soft Creamy Dessert

Milk (2% fat)	83.65
Sugar	12.00
Cocoa	2.70
Corn Starch*	1.50
Viscarin® GP 109*	0.15

Dessert Cream

Milk	81.95
Sucrose	13.00
Skimmed Milk Powder	2.00
Modified Starch	2.20
Carrageenan or Carrageenan/Xanthan Gum	0.40
Flavor and Color	as required

Tofu-Type Frozen Dessert

Formula No. 1

(6% Fat Soft Serve)

Water	65.70
Sugar	11.97
Maltrin® M250 Corn Syrup Solids	9.90
Soybean Oil	6.00
Tofu	3.30
Pro-Fam® S798 Isolated Soy Protein	2.50
Salt	0.30
Locust Bean Gum	0.16
Lecithin (dry)	0.10
Cellulose Gum	0.05
Vanilla	0.02

Blend all dry ingredients and add to the water. Begin heating. In a blender, mix the soybean oil and tofu. Add the soybean oil/tofu mixture to the

* Carrageenan starch rations must be balanced to adjust texture and other desired properties.

other ingredients when the temperature reaches 100 F. Pasteurize at 160 F for 30 min. Homogenize hot mix using a two-stage homogenizer: 1st stage—1500 psi; 2nd stage—500 psi. Cool, add flavorings, then freeze with a 90–100% overrun.

Recommended Flavoring Systems:

Maple Nut Flavored Product—After cooling, add 8.3 oz of Schilling/McCormick artificial maple flavoring for every 100 lb of mix. Add walnut chips or other nuts to desired appearance.

Chocolate Flavored Product—For a chocolate flavored product, replace half of the **Maltrin® M250** corn syrup solids and 10% of the soybean oil with 2.5% powdered cocoa and 3.1% sugar.

Vanilla Flavored Product—To achieve a vanilla flavored product, add a liquid vanilla concentrate to desired taste.

Additional Flavoring Suggestions—Add coffee or butterscotch flavoring to desired taste.

Note: This product has the benefits of high protein content, no cholesterol, and a smooth, creamy texture similar to that of frozen dairy desserts. Also because this product contains no dairy ingredients, it can be labeled as parevine.

No. 2

(10% Fat Hard Pack)

Water	61.20
Sugar	11.92
Soybean Oil	10.00
Maltrin® M250 Corn Syrup Solids	9.90
Tofu	3.30
Pro-Fam® S798 Isolated Soy Protein	3.00
Salt	0.30
Locust Bean Gum	0.20
Lecithin (dry)	0.10
Cellulose Gum	0.06
Vanillin Powder	0.02

Blend all dry ingredients and add to the water. Begin heating. In a blender, mix the soybean oil and tofu. Add the soybean oil/tofu mixture to the other ingredients when the water temperature of dry ingredients/water reaches 100 F. Pasteurize at 160 F for 30 min. Homogenize hot mix using a two-stage homogenizer: 1st stage—1500 psi; 2nd stage—500 psi. Cool, add flavorings, then freeze with a 90–100% overrun.

Recommended Flavoring Systems:

Maple Nut Flavored Product—After cooling, add 8.3 oz of Schilling/McCormick artificial maple flavoring for every 100 lb of mix. Add walnut chips or other nuts to desired appearance.

Chocolate Flavored Product—For a chocolate flavored product, replace half of the **Maltrin® M250** corn syrup solids and 6% of the soybean oil with 2.5% powdered cocoa and 3.1% sugar.

Vanilla Flavored Product—To achieve a vanilla flavored product, add a liquid vanilla concentrate to desired taste.

Additional Flavoring Suggestions—Add coffee or butterscotch flavoring to desired taste.

Note: To provide a quicker meltdown, decrease the **Maltrin® M250** content until desired meltdown is achieved.

Tofu listed in this formulation is a commercial tofu having a solids level of 15.2%. A dry tofu may be substituted for this commercial tofu at a usage level of 0.5% with the addition of 2.8% of water.

Papaya Cake

Pillsbury Plus Yellow Cake Mix (18$^1/_2$ oz)	1	pkg.
Papaya Puree	1	cup
Eggs	3	
Wesson Oil	$^1/_4$	cup
Triple Sec or Cream Sherry	$^1/_8$	cup
Nutmeg or Mace	1	tsp.

Preheat oven to 350 F. Grease 10 in. tube pan with **Crisco** or margarine. Do not use oil or aerosol shortening spray. Dust greased pan with flour to cover; set aside. Place cake mix, papaya puree, eggs, oil, triple sec and nutmeg (mace) into mixing bowl. Mix on low speed to moisten. Beat 2 min on high speed. Pour into pan. Bake 35–45 min. Cake will be a light golden brown in color; pulled at sides, insert toothpick (cake tester) in center until it comes out clean and dry. Cool cake on cooling rack for 10 min before removing from pan. Store with loosely covered Saran Wrap.

Walnut Cookie

A	Sugar	7 lb 4 oz
	Salt	1 oz
	Soda	2 oz
	Cinnamon	4 oz
	Margarine	2 lb 4 oz
	Hydrogenated Shortening	1 lb 13 oz
B	Whole Eggs	2 lb 4 oz
	Molasses	8 oz
	Water	8 oz
	Vanilla	1 oz
C	Cake Flour	8 lb 4 oz
	Medium Walnut Pieces	1 lb 12 oz

Cream all ingredients in Part A until smooth. Add Part B ingredients slowly and cream light. Add flour. Mix smooth. Fold in walnuts. Bake at 375 F until done. Deposit weight: $1^{1}/_{2}$ oz.

Walnut Muffin

A	White Flour	3 lb
	Milk	2 lb 1 oz
	Compressed Yeast	1 oz
B	Sugar	12 oz
	Butter	1 lb
	Whole Eggs	10 oz
	Egg Yolk	1 oz
	Mace	$^{1}/_{4}$ oz
C	Walnuts	10 oz

In Part A set sponge, allow to rise 1 h. Cream Part B together. Add eggs slowly. When sponge is ready, combine two mixtures and beat thoroughly. Mix Part C. Put in well-greased muffin tins. Allow to rest 15 min. Bake at 400 F for 25–30 min.

Doughnut Glaze

Powdered Sugar	100.00
Salt	0.25
Water	30.00
Methocel® K3 Premium	0.10

Dissolve **Methocel** in water and bring to boil. Blend in rest of ingredients (slowly). Temper 15 min at 120 F. Apply at 120 F to doughnut hot from the fryer.

Note: If so desired, **Methocel** product may be diluted in a carrier such as those used in commercially available standardized icing stabilizers.

Dry-Mix Cheesecake Base

Whiptreme	25.64
Chez Tone 615B	23.08
Sucrose	20.51
Sweet Buttermilk Powder	19.23
Dextrose	8.53
Viscarin® SD 389 Carrageenan	2.05
Tetrasodium Pyrophosphate	0.96
Color and Flavor	to suit

Preparation: Pour $1^1/_2$ cups (355 ml) cold milk into a small mixing bowl. Add cheesecake mix and blend at low speed with an electric mixer for 30 s. Beat at medium speed for 3 min. Pour into prepared crust or mold and refriegerate for 1 h.

Dry-Mix Low Sugar Imitation Jelly

Gelcarin® SA 426L Carrageenan	18.15
Adipic Acid	2.82
Sodium Benzoate	2.02
Dextrose	77.01

Disperse the blended dry ingredients by spoon stirring into a large saucepan containing 1 l of fruit juice. Heat the mixture while continuously stirring over medium heat until it boils; hold for 1 min at a boil. Pour into jelly jars and cap with tight fitting lids while solution is hot. Invert jars to insure pasteurization of caps. The spread may be stored unopened at room temperature—refrigerate after opening.

Note: Maltodextrin may be used instead of dextrose for standardization.

Long-Life Fruit Jelly

Water	76.10
Sucrose	15.00
Fruit Pulp	6.00
Citric Acid	0.40
Modified Starch	1.00
Carrageenan	0.80
Buffer	0.40
Flavor and Color	as required

Reduced-Calorie, Imitation Jam and Jelly Spreads

		Formula No. 1	No. 2
		(Jelly)	*(Jam)*
A	Water	39.00	—
	Fruit Juice Concentrate	10.50	—
	Semithawed Fruit	—	53.50
	Sucrose	25.00	—
	High Fructose Corn Syrup (55DE)	—	23.20
	Potassium Sorbate	0.10	0.10
	Potassium Citrate	—	0.20
B	Water	24.40	22.00
	Gelcarin® SA 536 Carrageenan	1.00	—
	Gelcarin® SA 329B Carrageenan	—	1.00
	Coloring	to suit	to suit

Combine ingredients listed in Part A in a kettle with continuous mild agitation. Heat to 70–75 C. Disperse the carrageenan into the water in Part B using continuous agitation. Heat this slurry to 80–85 C. Add hot solubilized carrageenan to Part A. Adjust the pH of the hot solution to 3.5–3.8 using a 50% citric acid solution. Fill and cap containers at a minimum of 66 C.

Capped containers must be quickly cooled to minimize the hydrolysis of the carrageenan in the finished product and to optimize the suspension of fruit pieces in the imitation preserve spreads.

Notes: Increasing the recommended carrageenan levels will allow the substitution of the sucrose or fructose with aspartame and/or saccaharin. If cloudiness can be tolerated other nonnutritive thickeners, such as guar gum or microcrystalline cellulose, may be added to the formulations to build the solids content and to reduce syneresis.

Due to acid variability in each fruit system, pretesting is required to determine the quantity of citric acid solution needed to obtain a specified pH.

These reduced sugar spreads have a full body and mouthfeel which is comparable to their high sugar (65° brix) counterparts but with less than half the calories.

Experimental Fruit Spread

A	4 × 1 Strawberries	48.8
	Rezista® Starch	2.6
	Neto® 7350 Corn Syrup	29.8
B	**Isosweet® 180** or **5500** HFCS	17.9
	Mira-Gel™ 463 Starch	0.9

Mix and cook ingredients in Part A to 217 F in a steam-jacketed kettle. Check solids with refractometer and adjust to 62% by adding water or cooking as necessary to evaporate excess moisture. To prepare Part B ingredients, mix the **463 Starch** into the HFCS to form a smooth syrup concentrate. Add to the hot ingredients (Part A) and blend. Hot fill and pasteurize to 180 F center temperature if necessary.

Fruit Topping

Fruit*	50.0
Sucrose	30.0
Water**	19.5
Viscarin® SA 439	0.3–0.5
Sodium Citrate (pH 3.8–4.0)	0.2

* Strawberries (fresh or frozen) or other fruit sweetened four parts fruit to one part sucrose

** Adjust for total solids of 40–45%.

Combine dry ingredients and disperse into thawed fruit and water. Agitate until dry ingredients are incorporated. HTST process at 90 C (194 F) for 10 s or other acceptable process may be used. Cool to 29 C (85 F) or below depending on use and filling methods.

Instant Fruit Pie Filling

I.Q.F. Fruit	43.94
Water	32.40
Sugar (fine granular)	20.00
Mira-Thik™ 468 Starch	3.27
Vegetable Oil	0.13
Citric Acid	0.13
Salt	0.13
Color	to suit

Premix: Mix oil into one half of sugar. Add other dry ingredients (except remaining sugar) and blend uniformly into a dry premix.

Pour one cup of water into mixer bowl at slow speed. Add the premix and mix until smooth. Add remaining sugar and finish mixing until smooth and glossy. Fold in fruit. Fill pastry shell and bake.

10% Fruit Filling for Pastry

Avicel® RC-591F MCC		1.250
Water		42.916
Raspberries		10.000
Sucrose		32.000
FD&C Red 40 (4%)		0.990
Purple #5857 (4%)		0.129
Caramel Coloring (20%)		0.225
Sodium Benzoate		0.050
Potassium Sorbate		0.050
Modified Starch }	starch	3.750
Water }	slurry	7.500
Imitation Flavor		0.020
Citric Acid		0.750
Sodium Citrate		0.370

Disperse **Avicel RC-591F** into hot water with vigorous agitation and mix until smooth and viscous (5–10 min). Continue heating. Add raspberries, all colorings, sodium benzoate, and potassium sorbate and mix well. Add starch slurry at 190–200 F (88–94 C). Maintain temperature and agitation for 10 min. Product will thicken at this point. Add flavor, citric acid, and sodium citrate. Fill into containers and cool. Deposit on or in pastry. Bake pastry and freeze. Defrost in microwave oven for 2–2$^1/_4$ min.

Cherry Pie Filling

Juice/Water	25.0
Sucrose and/or HFCS*	21.0
Starch (modified waxy maize)**	2.5
Methocel® K100M Premium	0.3
42 D.E. Corn Syrup	0.7
Fruit (frozen cherries, 5 + 1 pack)	50.0

Combine approximately $^2/_3$ of the juice and/or water with the sugars under agitation and heat to approximately 200 F. Slowly disperse dry **Methocel** powder into hot sugar solution under moderate agitation. Slurry starch in remainder of juice and/or water. Add starch slurry to above mixture. Heat back to 200 F and hold approximately 10 min. Add filling to cold fruit and gently mix in fruit pieces.

Note: Recipe can be used as a starting formula for all fruit pies, including berry, apple, peach, etc. Minor adjustments may need to be made in levels of starch and/or **Methocel** to help achieve your desired consistency. Minor adjustments in sugar levels may also be desired depending on the required sweetness. Sucrose, corn syrups, and high fructose corn syrups may all be used in varying combinations.

Lemon Pastry Filling

Fructose Corn Syrup 55	29.0000
Freezist M	5.0000
Methocel® K100M Premium	0.1250

* 42% high fructose corn syrup.

** Suggested products include **Col Fol 67**, **Thermflo**, or **Polar Gel 15**.

Guar Gum	0.0625
Locust Bean Gum	0.0625
Sodium Citrate	0.1900
Citric Acid (50% sol'n.)	0.5600
Yellow #5 Sol'n.	0.2500
Titanium Dioxide	0.0100
Sodium Benzoate	0.0500
Potassium Sorbate	0.0500
Water	48.5600
Water	16.0000
Lemon Oil	0.8000
BHA	0.0004

Add **Freezist M** to water. Add potassium sorbate and sodium benzoate to water. Add preblended gums (guar, locust bean gum) to starch slurry under moderate agitation, approximately 10 F. Add Fructose Corn Syrup 55. Add sodium citrate, titanium dioxide, and color solution. Heat to approximately 175 F. Add dry **Methocel** powder under moderate agitation. Add starch slurry and citric acid solution. Heat to 200–205 F for 15 min. Add BHA to lemon oil and dissolve in oil. After 15 min take sample of filling and check soluble solids and pH. Adjust solids if necessary. Cool filling to 180 F. Add lemon oil with BHA or BHT to filling. Pack, freeze.

Note: This filling gives the good clean cut-off properties important in pastry filling operations.

Lemon Pie Filling

Isosweet® 5500 HFCS	18.6
Lemon Oil	0.6
Yellow #5 (1%) Coloring	0.1
Mira-Gel™ 463 Starch	6.2
Citric Acid Crystals	0.2
Granulated Sugar	12.4
Cold Tap Water	61.9

Use Sunbeam Mix Master. Add syrup, coloring, and flavor to a clean, dry two-quart bowl and mix. Hand stir starch and citric acid into the syrup to avoid dusting. Then at slow speed, power mix into a smooth syrup. Continue mixing, and add water, *pouring it quickly*, into syrup. Scrape bowl sides with a spatula to insure thorough mixing of the water and syrup concentrate. Add sugar and complete mixing. Pour into pie shell. Slice and serve in 20–30 min.

Pumpkin Pie Filling

Formula No. 1

Strained Pumpkin Puree	62.00
Water	18.88
Sugar	9.80
Nonfat Dry Milk Solids	5.40
Eggs (frozen)	2.75
Spices	0.45
SeaGel® GP713* Carraggenan	0.50
Salt	0.22

Thoroughly blend sugar, milk solids, salt, spices, and **SeaGel® GP713.** Slowly disperse the above ingredients into the water and eggs in a slurry tank with constant agitation. Heat the strained pumpkin puree to 72 C (160 F) in a scraped surface kettle. Pump the water/egg slurry into the kettle containing the pumpkin puree and slowly heat, under constant agitation to 84 C (180 F). Hold at 84 C (180 F) for 5 min. Fill pie shells and bake at 450 F for 10 min, then 350 F for 50 min. Alternatively, the pie filling may be canned and retorted.

No. 2

Sugar	12.70
Powdered Milk	5.10
Methocel® K100M Premium	0.30
Amaizo W-13 Modified Waxy Maize Starch	1.30
Salt	0.32
Cinnamon	0.36
Nutmeg	0.12
42% High Fructose Corn Syrup	5.10
Pumpkin	34.20
Water	30.40
Whole Eggs	10.10

* Alternative stabilizer: **SeaKem® CM 614** carrageenan @ 0.30–0.60%.

Blend together dry ingredients such as sugar, powdered milk, **Methocel**, starch, salt, cinnamon, and nutmeg. Add dry ingredients to pumpkin and mix well. Add water and mix until smooth. Allow to stand at least 1 h at room temperature. Add eggs, mix well and fill pie shells. Bake pie at 425 F for 25 min.

Custard Pie Filling

Water	46.50
42 D.E. Corn Syrup	5.80
Sugar	17.40
Powdered Milk	4.40
Salt	0.18
Amaizo W-13 Modified Waxy Maize Starch	0.73
Butter	1.09
Vanilla	0.36
Egg Whites*	17.40
Egg Yolks**	5.80
Methocel® K100M Premium	0.20

Heat $^1/_4$ of water, corn syrup, and starch to boiling. Add the remaining water (cold). Dry blend the powdered milk, sugar, and **Methocel** and add to mixture. Add eggs, vanilla, salt, and butter (melted) and blend well. Fill pie shell $^1/_2$ full and bake 10 min or until crust forms. Refill as desired and bake approx. 15 min more.

Note: This formula represents a 50% reduction in egg yolks as compared with typically used formulas.

Custard Type Dry Mix

(with Water)

	g/Package
SeaGel® FL Type Carrageenan	1.00–2.75
Cane Sugar	55.00
Nonfat Milk Solids	50.00

---

* 1 pt. Henningsen Foods, Inc. Egg white solids spray-Type P20 and 7 pts. water.

** 1 pt. Henningsen Foods, Inc. Egg yolk solids Standard-Type Y-1 and 1.25 pts. water.

Spray Dried Emulsified Fat	10.00
Cocoa	7.00
Salt	0.30
Sodium Hexametaphosphate	0.50

Add dry mix to 2 cups (475 ml) of water in a saucepan and stir unti thoroughly blended. Cook and stir over medium heat for approximately 5 min until mixture comes to a boil. Remove from heat, pour into mold and allow to cool quiescently at room temperature or in a refrigerator.

Demoldable Custard

Formula No. 1

Milk	85.20
Sucrose	12.00
Modified Starch	2.00
Carrageenan	0.30
Flavor and Color	as required

No. 2

(Dietetic)

Skim Milk	97.00
Sweetener	0.10
Modified Starch	1.80
Carrageenan	0.40
Flavor and Color	as required

Carrageenan-Based Custard-Like Puddings

	Formula No. 1	No. 2	No. 3 *(Reduced Calorie)*
Fluid Whole Milk	78.00	80.02	95.76
Sucrose	15.39	12.47	—
Cocoa	2.56	2.67	3.14

Modified Waxy Maize Starch	2.56	—	—
Whiptreme	0.99	4.06	—
Disodium Phosphate	0.30	0.38	0.46
Gelcarin® GP359 Carrageenan	0.16	0.25	0.33
SeaKem® CM 614 Carrageenan	—	0.10	0.12
Salt	0.04	0.05	0.06
Saccharin and/or Aspartame	—	—	0.13
Flavor and Color	to suit	to suit	to suit

Premix dry ingredients and disperse into cold milk with agitation. Preheat to 66–71 C (150–160 F) in a steam jacketed kettle with constant agitation. An HTST process is now employed. A custard-like set can be obtained by filling at temperatures above the gelling point > 54 C (> 130 F) of the formulations. This avoids fracturing the gel during filling. A pudding-like texture can be obtained by filling at 27–38 C (80–100 F) which is below the gelling point. The starch in formula No. 1 also contributes to a more puddling-like texture.

Note: If nonfat milk solids, skim milk, or condensed milk are used in place of fluid whole milk, the pudding should be homogenized prior to sterilization to disperse the additional fat.

Dry-Mix Cold Process Pudding

Sucrose	67.00
Dextrose	26.00
Lactarin MV 406* Carrageenan	3.00
Tetrasodium Pyrophosphate	2.00
Disodium Phosphate	1.00
Sodium Chloride	0.50
Vegetable Oil	0.50
Color and Flavor	to suit

Pour 1 pint of cold (10 C) whole milk into a mixing bowl. Add dry powder mix to milk while stirring with wire whisk for $^1/_2$ min. Pour into container and refrigerate for 10–15 min. Pudding is ready to serve.

* Alternate carrageenan: **Lactarin XP 3040**

Vanilla Pudding

Insta Starch 5/70	83.300
Tetrasodium Pyrophosphate	7.300
Disodium Phosphate	1.500
Artificial Vanilla Flavor	1.450
Yellow #5	0.046
Yellow #6	0.004
Flour Salt	1.800
Aspartame	1.600
Mono and Diglycerides **G695**	3.000

Blend flour salt and 10% of the **Insta Starch** with melted **G695** mono and diglyceride mixture. Blend with remaining ingredients in a V blender.

To mix: Add 32.5 g of pudding mix to 2 cups of cold milk and mix for 1 min with a wire whip or electric mixer on low speed. Pour into cups and refrigerate for 9 min.

Frozen Yogurt

A	Water	14.21
	Sucrose	13.70
	36 DE Corn Syrup	5.00
	Avicel® RC-591	0.35
	Tween® Mos. 100K	0.10
	Guar Gum	0.19
	Carrageenan (LMR)	0.02
	CMC-7M8SF	0.13
B	Water	49.10
	MSNF	10.80
	Butterfat	2.40
	Yogurt Culture (12% solid)	4.00

Stabilizer and sweetener added to cultured yogurt.

Frozen Confection Fudge Bar

	Formula No. 1	No. 2
Cocoa	2.00	2.00
MSNF	13.00	6.50
Whey	—	6.50
Sucrose	14.00	13.90
42 DE Corn Syrup	4.00	3.95
Stabilizer	0.55–0.75	0.75–1.25
Water	to yield 33.8% total solids mix	

This stabilizer will yield a superior product even with the substitution of 50% whey solids for milk solids.

Stabilizer Blend:

Tween® Mos. 100K	17.46
Guar (TIC)	37.31
Avicel® MCC RC-581	42.48
Carrageenan (LMR)	2.75

Dry Ice Cream Mix

Butterfat	3.50
MSNF	15.00
Sucrose	10.00
Corn Syrup Solids	4.50
Avicel® RC-591	0.80
CMC-7HP	0.15
Polysorbate 80	0.10

Dry-Mix Water Dessert Gels

	Formula No. 1	No. 2
Sucrose	93.37	93.20
Gelcarin® DG 425B	3.53	—
Gelcarin® DG 437B	—	4.70
Adipic Acid	2.10	2.10
Tripotassium Citrate	1.00	—
Flavor and Color	to suit	to suit

Pour 500 ml of boiling water into a mixing bowl containing the dry mix. Stir for 2 min or until fully dissolved. Pour the solution into a mold and allow to set quiescently at refrigerated or room temperature. Gels will be ready to consume in 1–2 h.

Note: Adipic acid has been used because of its very low hygroscopic property, making it especially suited for package dry mixes. Fumaric acid may effectively be used with formulation modification regarding pH adjustment.

Walnut Kisses

Egg Whites	5 lb
Granulated Sugar	5 lb
Salt	2 oz

Heat to 105 F on double boiler, stirring constantly. Beat with wire whisk to medium peak; by hand, fold in very lightly (to avoid overmix) 7$^1/_2$ lb 6 × granulated sugar, 3 lb of toasted cake crumbs (run through sieve); 5 lb of ground walnuts and 2 oz vanilla. Bag to size of quarter. Bake in cool oven at 325 F until lightly colored.

Candy Tablets

(Nonplaque Producer)

Icing Sugar	100.000
Stearic Acid	0.750
Isopropanol	1.250
Gelatin	0.750
Glucose Syrup	0.500
Water	4.000
Flavoring	0.125
Calcium Lactate	5.000

Sulfiting Replacement in Foods

	Formula No. 1	No. 2	No. 3
Citric Acid	90	80	60
Ascorbic Acid	10	20	40

Note: The most appropriate blend will depend on the specific needs of a particular situation. If the conditions under which the fruits and vegetables are processed and stored are not the best, then a blend with a higher ratio of ascorbic acid to citric acid will be required.

Each of these combinations can be used by dissolving 1 oz of the blend per gallon of cold water. Chopped vegetables or fruits are then dipped in this solution for 30–45 s, spin dried, and stored in closed polyethylene bags, under refrigeration, until used. Under these conditions the fruits and vegetables should retain their natural, fresh appearance and flavor for up to 5 days after being processed.

Chapter IV

COSMETICS

Deodorant Stick

Syncrowax AWI-C	7.2
Lanexol AWS	1.3
Propylene Glycol	68.0
Sodium Hydroxide	0.8
Irgasan DP 300	0.1
Dye, Perfume	q.s.
Water	22.6

Heat **Syncrowax AWI-C, Lanexol AWS**, and propylene glycol to 75 C. Add a solution of the sodium hydroxide in the water which is also at 75 C. Mix and raise temperature to 80–85 C. Mix well, cool to 70 C, add **Irgasan** and pour into case.

Antiperspirant Stick

Formula No. 1

Syncrowax HGL-C	12.0
Syncrowax HRC	3.0
Stearic Acid (triple pressed)	3.0
Crodamol PMP	44.5
Talc	17.0
Microdry	20.0
Cab-O-Sil® M5	0.5

Combine and heat to 80 C with agitation. Ensure all waxes are melted and powders are completely dispersed. Cool to 45 C and fill.

No. 2

Rehydrol®	20.0
Propylene Glycol	26.0
Monamid® S	26.0
Isocetyl Alcohol	11.3
SDA 39C (anhyd. ethanol)	14.5
Isopropyl Myristate	2.0
Titanium Dioxide	0.2
Fragrance	q.s.

Add **Rehydrol®** slowly with stirring to propylene glycol. Heat to 80 C; attain solution. Add **Monamid® S** slowly until dissolved maintaining temperature at 80 C. Add isocetyl alcohol and allow solution to cool to 76 C; then add ethanol, then isopropyl myristate. Add titanium dioxide and fragrance and continue stirring until resultant suspension is homogeneous. Slow stirring rate, allow to cool to 70–72 C and pour into stick casing covering the stick with vinyl wrap. Remove air pocket by pressing down slightly on vinyl wrap. This prevents holes from forming in the stock due to entrapped air.

No. 3

A	**Veegum® HV**	1
	Water	20
	Aluminum Chlorhydrate (impalpable)	18
B	**Volatile Silicone 7207**	26
	Isopropyl Myristate	5
	Stearyl Alcohol	24
	Arlacel® 85	5
	Tween® 85	1
	Preservative	q.s.

Slowly add **Veegum HV** to the water, while agitating at maximum available shear. Continue mixing until smooth. Add aluminum chlorhydrate and mix until smooth. Heat Part A with stirring to 70 C. Heat Part B with stirring to 65 C. Add Part A to Part B. Mix to 50 C. Pour into molds and cool.

Consistency: Firm stick.
Suggested Packaging: Push or twist-up dispenser.

No. 4

Stearyl Alcohol	26.00
2-Ethylhexyl Palmitate	14.50
Dioctyl Adipate	14.25
Volatile Silicone	20.00
Aluminum Zirconium Glycine (super fine powder)	20.00
Arachidyl Propionate Polymer (powder)	5.00
Wheat Germ Glycerides	0.25
Fragrance	q.s.

No. 5

Stearic Acid Monoethanolamide	27.75
2-Ethylhexyl Palmitate	22.00
Dioctyl Adipate	25.00
Aluminum Zirconium Glycine (super fine powder)	20.00
Arachidyl Propionate Polymer (powder)	5.00
Wheat Germ Glycerides	0.25

No. 6

Syncrowax HR-C	2.0
Syncrowax HGL-C	4.0
Crodacol S (USP)	24.0
Crodamol PMP	43.0
Prostearyl 15	5.0
Microdry Ultrafine	20.0
Cab-O-Sil® M5	1.0
Volpo 5	1.0
Perfume	q.s.

Heat **Crodamol PMP** and **Prostearyl 15** to 70 C. Add **Crodacol S, Syncrowax, Cab-O-Sil, Volpo** with stirring. Then add **Microdry**; when uniform, cool to 55 C. Add perfume and fill.

No. 7

Syncrowax HR-C	5.0
Volatile Silicone 7158	41.8
Crodamol PMP	10.0
Microdry Ultrafine	20.0
Crodacol S (95% USP)	20.5
Cab-O-Sil® M5	1.0
Volpo 3	1.0
Verbena 40188 B	0.7

Roll-On Antiperspirant

Formula No. 1

(Dry)

Volatile Silicone	40.00
Bentone® 38 Powder	2.00
SDA-40 Ethanol (200 proof)	2.00
Volatile Silicone	32.90
Isopropyl Myristate	2.00
Wheat Germ Glycerides	0.10
Aluminum Zirconium Glycine (super fine powder)	20.00
Demineralized Water	1.00

No. 2

Aluminum Sesquichlorohydrate Beads	20.0
Propylene Glycol (USP)	10.3
Glycine	0.2
Demineralized Water	21.9
Brij® 30 SP	2.0
SDA-40 Ethanol (190 proof)	44.0
Dioctyl Adipate	1.5
Wheat Germ Glycerides	0.1

No. 3

(Lotion)

A	**Veegum®**	1
	Water	51
B	**Arlacel® 165**	8
C	Aluminum Chlorhydrate (50%)	40
	Preservative	q.s.

Slowly add **Veegum** to the water, while agitating at maximum available shear. Continue mixing until smooth. Heat Part A to 70 C. Heat Part B to 75 C. Add Part B to Part A and mix to 50 C. Heat Part C to 50 C and add to Parts A and B. Mix until cool.

Consistency: Medium viscosity lotion.

Suggested Packaging: Roll-on container.

Note: **Veegum** thickens, stabilizes, and minimizes tack in this roll-on. This formula is a good base for modification with additional ingredients to attain specific product properties.

No. 4

(Lotion)

A	**Veegum® HV**	1
	Water	49
B	**Amerchol L-101®**	5
	Solulan® 98	2
	Cetyl Alcohol	1
	Glycerin	2
	Arlacel® 165	4
C	Aluminum Chlorhydrate (50%)	36
	Preservative	q.s.

Slowly add **Veegum HV** to the water, while agitating at maximum available shear. Continue mixing until smooth. Heat Part A to 85 C. Heat Part B to 80 C. Add Part A to B with mixing to 40 C. Heat Part C to 40 C and add to Parts A/B slowly. Continue stirring and cool to room temperature.

Consistency: Medium viscosity lotion.

Suggested Packaging: Roll-on container.

Note: **Veegum HV** thickens and stabilizes this tack-free roll-on.

No. 5

(Lotion)

A	**Arlamol® E**	4.5
	Brij® 721	1.0
	Brij® 72	4.5
B	Water	45.0
C	**Arlatone PQ220**	10.0
D	**Wickenol® 303** (50% sol'n.)	35.0

Heat Part A to 70 C and Part B to 72 C. Add Part B to A with efficient agitation. Cool to 60 C with stirring and add Part C. Cool to 50 C with stirring and add Part D. Continue mixing until uniform.

Physical Properties:

Typical Visc. — 3800 cps

Stability — 3 months @ 25 C and 40 C

No. 6

(Dry)

Crodamol PMP	34.0
Syncrowax HGL-C	4.0
Volatile Silicone 7207	34.0
Microdry Ultrafine	19.0
Procetyl AWS	7.5
Water	1.5

Mix **Crodamol PMP** and **Syncrowax HGL-C** and heat to 75 C. Disperse **Microdry.** Add **Volatile Silicone,** mix **Procetyl AWS** and the water, then add to oils, perfume, cool, and fill.

Low-Cost Antiperspirant

(With Corn Starch)

Crodamol PMP	15.0
Siloxane F222	10.0
Crodacol S-95	8.0
Syncrowax HGL-C	8.0
Syncrowax HR-C	4.0
Prostearyl 15	5.0
Volpo 5	1.0
Titanium Dioxide	1.0
Chlorohydrol Fine	20.0
Dry Flo	28.0
Fragrance	q.s.

Add waxes to batch kettle equipped with a condenser and melt at 70 C under controlled agitation. When waxes are completely melted add **PMP, Prostearyl,** and **Volpo,** maintaining temperature and agitation. Add **Siloxane.** When uniform, slowly add **Chlorohydrol,** titanium dioxide, and **Dry Flo** in order. Allow to mix for 2 h to uniformity. Cool to 66 C. Pour off.

Talc Spray Antiperspirant

Bentone® Gel IPM	8.0
SDA-40	2.0
Isopropyl Myristate	1.5
Cyclomethicone	5.0
Aluminum Chlorhydrate	6.0
Talc 141 B.C.	2.0
Fragrance	0.5
Propellant A-46	75.0

Perfume Fluff

(Low Alcohol)

A	**Satexlan 15**	0.50
	Satexlan 5	0.50
	Liquid Base CB3896	1.00
B	**Cellosize® QP4400**	0.15–0.30
	Deionized Water	71.85
	Croderol G7000	1.00
	Carbopol® 941 (1% aq. sol'n.)	8.00
C	Ethanol	15.00
	Perfume	1.00–3.00
D	Diisopropanolamine (10% sol'n.)	1.00–1.50 to pH 6.5–7.5

Heat Part A to 70 C. Blend Part B and heat to 70 C. Add Part B to Part A under high-speed mixer. Add Part C when temperature falls to 60 C. Finally add Part D at 50–55 C and stir until cool.

Pack:	Concentrate	94.0
	Arcton 12/114 (60:40)	6.0

Solid Perfume

Paraffin Wax (140 F)	5.0
Syncrowax ERLC	10.0
Syncrowax HGLC	10.0
Perfume	10.0–20.0
Crodamol IPP	55.0–65.0

Gently heat the waxes with the **Crodamol IPP**. The perfume is added before the base solidifies, at the coolest temperature possible to avoid distortion. Fill off.

Note: The cost of the above product may be reduced by a partial replacement of the **Crodamol IPP** by mineral oil.

Perfume Stick

Formula No. 1

(Nonalcoholic)

A	Propylene Glycol	60.0
	Water	15.0
	Standamul® O-20	10.0
	Standamid® LD	5.0
	Sodium Stearate (NF)	9.0
B	Perfume Oil	q.s.
	Dyes and Preservatives	q.s.

Blend Part A thoroughly at 90 C. Cool to 60 C and add Part B with constant agitation. Blend until uniform and at 60 C fill into molds.

Note: The ethoxylated fatty alcohol and amide contribute to the structure and clarity of this stick. As much as 5% perfume oil may be used in this system.

No. 2

Duron 0423	5.5
Ozokerite 71/74	14.0
Coconut Oil (110 C)	30.0
Glucate® SS	7.5
Magnesium Carbonate	2.0
Amerchol L 101®	26.0
Perfume Oil	15.0
Preservative, Color, and Antioxidant	9.5

Heat all ingredients except perfume oil to 85–89 C; add fragrance at 75 C. Keep warm while mixing to uniform paste. Mold into sticks.

Cream Perfume

Formula No. 1

A	Perfume	4.0–8.0
	GMS S/E	2.00
	Crosterene SA4130	14.00

B	**Liquid Base CB3896**	3.0
	Preservative	0.2
C	**Croderol G7000**	4.0
	Potassium Hydroxide	0.5
	Deionized Water	76.3

Heat Part B to 65–70 C. Heat Part C to 65–70 C and add to Part B under high-speed mixer. Add Part A to 50–55 C and stir until cream begins to thicken. Allow to stand then restir. A soft cream forms on standing.

No. 2

(Emollient)

A	Cetyl Alcohol	5.0
	Crill 3	2.5
	Crillet 3	3.0
	Corona Lanolin BP	1.0
	Mineral Oil	6.0
	Preservative	0.2
B	Sorbitol (70% sol'n.)	2.5
	Deionized Water	71.9
C	Perfume	4.0–8.0

Heat Part A to 65–70 C. Heat Part B and add to Part A slowly under high-speed mixer. Add perfume at 40–60 C and stir until cream thickens.

Roll-On Perfume

A	**Klucel® HE** (1.75% aq. sol'n.)	42.0
B	SDA-3 Alcohol	48.0
	Pationic ISL	5.0
	Perfume HST C-77-7	5.0

Prepare Part A at room temperature and stir until clear. Combine ingredients of Part B and add Part B to A with agitation. Continue stirring until clear and homogeneous. Filter if desirable.

Physical Properties:

pH — 6.2
Visc. (Brookfield RVT spindle #2, 5 rpm @ 80 F)— 1900 cps
Cloud Pt. — 4 C

High-Fragrance Moisturizing Cologne-Type Skin Balms

	Formula No. 1	No. 2	No. 3
	(Alcoholic Emulsions)		
Volpo 3	1.0	—	—
Volpo 10	—	3.0	1.5
Polychol 5	—	—	1.5
Crodafos N-10 (neutral)	1.0	—	—
Crodamol IPM	—	5.0	5.0
Perfume	2.0	1.0	1.0
SD-40 Ethanol (95%)	40.0	33.0	33.0
Carbopol® 941	0.4	0.3	0.3
Triethanolamine	0.4	0.3	0.3
Water	55.1	57.4	57.4

Dissolve the emulsifiers and amine and oil in the alcohol. Disperse the **Carbopol** in water until completely dispersed. (Note: It may be desirable to make a 2% stock solution of **Carbopol.**) Mix the two phases together.

Notes: Inclusion of small quantities of the appropriate sequesterants and UV absorbers is useful in preventing light and trace metal catalyzed depolymerization of the **Carbopol** and the loss of viscosity which results. Care in neutralizing to just below neutrality is also helpful in emulsion stability. At the same alcohol level, dissolving the emulsifiers and oils in the ethanol followed by the addition of the water, **Carbopol**, and amine will result in less opaque emulsions than will the method given. Additional alcohol or perfume tends to destabilize these systems so care must be taken to rebalance the emulsion when changes are made in the oil phase ingredients.

Fluff-Type Moisturizing Cologne

(Aerosol)

Crodafos N.3 (neutral)	1.50
Polychol 5	3.00
Polawax A.31	1.25
Novol	2.00
SD-40 Ethanol (95%)	46.00
Perfume (1.5–5%)	q.s.
Isobutane/Propane	5.00
Water	41.25

Procedure is the same as for above formulas for alcoholic emulsion skin balms but it should be noted that the water/alcohol ratio is critical and may require adjustment depending on perfume and concentration.

Note: The lightness and stability of the foam puff may be improved, if needed, by the addition of 0.5% **Carbopol 941.**

To produce a stable foam it may be necessary to lower the concentration of alcohol in the formula because of the superior compatibility of hydrocarbons in hydroalcoholic systems.

Fill:	Concentrate	95
	Propellants	5

Fragrance Stick

Cyclochem 326 A	13.5
White Ozokerite 170D	20.0
Carnauba Wax Yellow No. 1	1.8
Isopropyl Myristate	42.0
Rice Wax (RW001-619)	4.0
Neobee M-20	6.3
Diisopropyl Adipate	7.2
Tenox 4	0.1
Propyl Paraben	0.1
Pefume Oil	5.0

Heat ingredients to 80 C. Add perfume oil and mix well. Pour into mold.

Fragrance Sachet

A	Water	42.8
	SD-40 Ethanol	45.0
	Spectra-Sorb UV284	0.1
	Hamp-ene 100	0.1
	Carbopol® 934	1.0
B	Triethanolamine (99%)	1.0
C	Glyceryl Monostearate	4.0
	Standamul® 318	5.0
	Eumulgin® B-1	1.0
D	Perfume Oil	q.s.
	Dyes and Preservatives	q.s.

Blend Part A in the order given at 50 C. Adjust the pH to 7.5 ± 0.5 with triethanolamine. Blend Part C at 65 C. Add Part C at 65 C to the combination of Parts A and B with efficient mixing. Cool and continue stirring to 40 C. Add individual components of Part D under agitation. Blend until uniform and fill at 40 C.

Note: The ethoxylated fatty alcohol emulsifier provides auxiliary surfactant properties in this o/w cream. As much as 15% perfume oil may be used in this system.

Perfumed Cream Sachet

A	**Lexemul® 561**	20.00
	Lexol GT 865	2.50
	Lexol PG 865	2.50
	Cetyl Alcohol	5.00
	Lexgard P	0.05
	Perfume	q.s. to desired level or 6.00
B	Poloxamer 105	2.00
	Propylene Glycol	2.00
	Lexgard M	0.15
	Lexgard P	0.05
	Water	q.s. to 100.00

Add ingredients of Part B except for perfume to water at 70 C. Separately melt contents of Part A together. Slowly add Part A to B with agitation. Cool to 35 C. Add perfume and fill.

Physical Properties:

pH — 4.0
Visc. (@ 24 C) — 155,000 cps
Stability after one month — Stable at room temperature, at 45 C, and passes 3 freeze/thaw cycles

Bubble Bath

Formula No. 1

Water	63.37
Ammonium Chloride	0.50
AOS-A246-LX	17.50
Akypo RLM-100	7.80
Akypo RLM-45N	1.00
Triethanolamine	1.83
Sipomide 1500	5.00
Glucamate® DOE-120	3.00
Fragrance	q.s.

Dissolve ammonium chloride in water. Add **AOS** and the two **Akypo** surfactants with adequate agitation to assure a uniform dispersion. Add triethanolamine with agitation to assure uniformity. Add **Sipomide** and **Glucamate DOE-120**. Warming the product will help solubilize the **Glucamate**. Add a suitable fragrance.

Physical Properties:

Visc. (@ 25 C) — 6300 cps
Solids — 25.23%
pH — 7.65
Appearance — Yellow clear liquid

No. 2

Distilled Water	34.0
Ammonium Chloride	1.0
Alconate L-3	20.0
Sipon® LT-6	30.0
Sipoteric GSC	10.0
Sipomide 843	5.0
Fragrance, Color, Preservative	q.s.

Dissolve ammonium chloride in water with moderate agitation. Add remaining ingredients, singly, with continued agitation to obtain a uniform dispersion before adding next ingredient.

Physical Properties:

Appearance — Clear liquid
pH — 7.8
Visc. (# 3 spindle, 6 rpm, 25 C) — 6900 cps

	No. 3	No. 4
Neodol® 25-3S (60% AM)	32.0	32.0
Ammonyx LO (30% AM)	16.0	16.0
Bubble Bath Fragrance No. 959	1.0	—
Shampoo Fragrance No. 3770	—	0.3
FD&C Blue No. 1 (1% aq. sol'n.)	0.1	0.1
Water	q.s.	q.s.*

	No. 5	No. 6	No. 7
Neodol® 25-3S (60% AM)	20.0	20.0	20.0
Monamine ADD-100	3.0	2.0	2.0
Methyl-*p*-Hydroxybenzoate	0.1	0.1	0.1
Citric Acid	0.3	0.3	0.3
Sodium Chloride	4.0	3.5	3.0
Water	72.6	74.1	74.6

* Quantity sufficient to make 100% total.

No. 8

A	**Quatrisoft™ Polymer LM-200**	1.0
	Water	46.5
B	Sodium Myreth Sulfate (58% act.)	38.0
	Lauramide DEA	7.0
	Glucam® E-20	3.5
	Solulan® L-575	4.0
	Perfume and Preservative	q.s.

Disperse the **Quatrisoft Polymer LM-200** with good agitation in water at room temperature. When thoroughly dispersed, heat to 45 C with continued mixing. When a clear, uniform solution has developed, begin to add the ingredients of Part B in the order in which they are listed, mixing moderately after each addition. Avoid air entrapment.

No. 9

Miranol® CS Conc.	25.0
Mirataine® CBS	10.0
Cedepon® LS 30PM	10.0
Super-Amide L-9C	1.0
Water	54.0

Add in order above. Heat to 60 C to dissolve the **Super-Amide L-9C.** Adjust pH to 7.0 with citric acid.

Physical Properties:

Solids — 20.2%
Visc. — 1800 cps

No. 10

(Concentrate)

Carsonol SLES (30%)	39.90
Carsonol ALS	37.50
Carsamide SAL-9	8.00

Lonzaine CO	5.00
Lonzest 143-S	1.50
Citric Acid	0.55
Water	7.55

Physical Properties:

pH — 5.0
Visc. — approx. 1700 cps

No. 11

(Baby, Mild)

Water	15.0
Monateric LMAB	25.0
Monamate LNT-40	30.0
Monateric 951A	30.0

Combine ingredients. Adjust pH to 6.0–6.5.

Physical Properties:

Appearance — Clear liquid
Nominal Act. — 27%
Visc. — 500 cps

Cream Bath

A	**Secosol ALL/40**	30.0
	Ninol 128 Extra	5.0
	Monthyle®	5.0
	Secosyl	10.0
	Amphosol AA	10.0
	Demineralized Water	34.1
B	Formaldehyde (30% sol'n.)	0.2
	Citric Acid (50% sol'n.)	0.2
	Vegetol® Camomile MCF 793 (water soluble)	2.5
	Vegetol® Broad Leaved Lime MCF 762	1.0
	Perfume (water soluble)	2.0

Part A is heated to 65–70 C. Cool while stirring slowly. At about 40 C, the other components are added.

Foam Bath

A	Demineralized Water	30.00
	Secosol ALL 40	26.00
	Secosyl	10.00
	Amphosol AA	11.00
	Ninol 128 Extra	4.00
B	Demineralized Water	7.55
	Sodium Chloride	2.00
	Citric Acid	0.10
	Formaldehyde (30% sol'n.)	0.25
	Vegetol® Camomile (water soluble)	0.50
	Vegetol® Marigold (water soluble)	0.50
	Vegeteol® Broad Leaved Lime (water soluble)	0.10
	Perfume (water soluble)	8.00

At room temperature, Part A and Part B are mixed. Add perfume as required.

Baby Bath

Formula No. 1

Water	65.9
Sipex® EST-30	23.3
Sipoteric COB	8.6
PEG-6000 Distearate	1.5
Benzyl Alcohol	0.1
Sodium Phosphate	0.2
Dowicil® 200	0.1
Disodium Phosphate	0.2
Tetrasodium EDTA	0.1
Fragrance	q.s.

Add **Sipex EST-30** and **Sipoteric COB** to water with moderate agitation while heating to 65 C. At 65 C add PEG-6000 distearate while continuing agitation. When solubility is obtained continue agitation and begin cooling. Add remaining ingredients singly with agitation to obtain solubility before making next addition. Add a suitable fragrance at 45–50 C.

Note: Viscosity may be increased by increasing the **Sipoteric COB.**

Physical Properties:

Appearance — Clear liquid
Visc. (@ 25 C) — 1660 cps
pH — 7.2
Solids — 13.3%

No. 2

Water	42.75
Methyl Paraben	0.15
Propyl Paraben	0.10
Propylene Glycol	0.50
PEG-1000 Monolaurate	2.00
Sipon® ESY	40.00
Sipoteric COB	14.30
Citric Acid	0.20
Fragrance	q.s.

Add water and begin heating to 80–85 C. Add the parabens and propylene glycol with moderate agitation. When solubility of parabens is obtained, discontinue heating and begin cooling with continued agitation. Add PEG-1000 monolaurate at about 65 C. Add remaining ingredients singly, agitating to obtain a uniform mixture before making next addition. Add a suitable fragrance at 45–50 C.

Note: Viscosity may be increased by addition of sodium chloride or reduced by increasing the PEG-1000 monolaurate.

Physical Properties:

Appearance — Clear liquid
Visc. (@ 25 C) — 190 cps
pH — 4.7
Solids — 17.95%

No. 3

A	Demineralized Water	25.0
	Secosol ALL 40	26.0
	Secosyl	10.0
	Amphosol AA	11.0
	Cocamide DEA	4.0
B	Demineralized Water	10.8
	Citric Acid	0.2
	Fondix® GBIS	1.0
	Vegetol® Marigold (water soluble)	3.0
	Vegetol® Camomile (water soluble)	3.0
	Sodium Chloride	2.0
C	Perfume	2.0
	Solvating Agent γ 2420	2.0

At room temperature, mix the components of Part A, and mix the components of Part B. Add Part B to Part A, then add perfume and solvating agent.

Shower Mousse

Copolymer 937	2.5
Distilled Water	40.5
Sodium Lauryl Ether Sulfate	40.0
Gafamide® CDD-518	2.0
Cocamidopropyl Betaine	10.0
Fragrance	q.s.
Propellant A-46	5.0

Dissolve **Copolymer** in water and stir until uniform. Add SLES with stirring. Add remaining ingredients. Put into container and pressurize with **A-46.**

Shower Gel

Formula No. 1

A	Demineralized Water	39.10
	Odorisol SPS	30.00
	Ninol 128 Extra	3.00
	Kathon CG	0.10
	Citric Acid (50% aq. sol'n.)	0.20
B	**Ateloglycane®**	3.00
	Demineralized Water	6.00
C	Perfume	0.50
	Solvating Agent γ 2428	1.00
D	**Amphosol CA**	7.00
	Sodium Chloride	1.50
	Demineralized Water	8.50
	Triethanolamine (50% aq. sol'n.)	0.10

Prepare the blend of Part A and add Part B to Part C. Add **Amphosol** and sodium chloride solution. Adjust pH to about 6.8–7.0 with triethanolamine, if necessary.

No. 2

(Clear)

Sodium Lauryl Sulfate	10.5
Amphoteric 18	6.0
Sodium Lauroyl Sarcosinate	3.0
Cocamide MEA	3.0
Water, Perfume, Color, Preservative	q.s. 100.0

Adjust to pH 6.2 with citric acid.

Bath and Shower Gel

A	**Standapol® ES-2**	51.00
	Velvetex® BK-35	10.00
	Deionized Water	26.90
	Ceraphyl 60	2.00
B	Sodium Hydroxide (25% aq. sol'n.)	q.s. to pH 6.5 ± 0.5
C	Sodium Chloride USP (20% aq. sol'n.)	10.00
	Formalin	0.10

Charge Part A into vessel and mix until uniform (at ambient temperature). Adjust pH to 6.5 ± 0.5 with Part B. Add Part C under slow agitation and package when uniform.

Bath Gel

Formula No. 1

Mirataine® CBS	29.0
Maprofix® ES	29.0
Witconate AOS	18.0
Ethyl Alcohol	3.0
Water	21.0

Mix all ingredients together and adjust pH to 7.0 with citric acid.

Physical Properties:
Solids — 30.1%
Visc. — 60,000 cps

No. 2

(Clear)

Sipon® ESY	35.0
Coco Monoisopropanolamide	8.0
Citric Acid (25% sol'n.)	0.5

Water	56.5
Preservatives	q.s.
Perfume	q.s.

Add **Sipon ESY** to water at 50 C. Add premelted amide under slow agitation and mix until homogeneous. Add citric acid, preservative, perfume, and color at 40–45 C.

Physical Properties:

Appearance — Clear gel
Cloud Point — 0 C
pH — 7.5
Solids — 17.5%
Color — Light yellow

No. 3

(At 23% Solids)

Sipoteric GSC	71
Water	29
Citric Acid (50% sol'n.)	q.s.
Perfume and Dye	q.s.

Heat water to 40–50 C. With agitation, add **Sipoteric GSC** to water. Heat at 40–50 C while stirring, until mixture is homogeneous. With agitation, add citric acid sol'n. Adjust pH to 4.5. Add perfume and dye as desired.

Physical Properties:

Appearance — Clear gel at room temperature
Solids — 22–24%
pH (10%) — 4.5–5.0

No. 4

(Nonirritating)

Water	30.8
Monateric CAB	23.3

Monamate OPA-30	25.9
Sipon® ESY	20.0

Add ingredients in the order listed while heating to 60 C. Add perfume, color, and pack while hot.

Note: Draize tests at 10% active dilution showed a very mild irritation of the conjunctivae, with clearing on the second day. The Primary Irritation Index was 0.5, indicating a very low degree of skin irritation.

Physical Properties:

pH as is — 6.2–6.4
Visc. — 30,000 cps
pH adjusted — 5.0
Visc. — 65,000 cps

Bath Oil

Formula No. 1

(Spreading)

Drakeol 7 (USP)	48.0
Isopropyl Myristate	46.0
Perfume	5.0
PEG 40 Sorbitan Peroleate	1.0

Blend all ingredients at room temperature.

Note: These oils form a very thin film which floats on the surface of the water.

No. 2

(Dispersible)

Drakeol 7 (USP)	65.0
Isopropyl Myristate	20.0
PPG-15 Stearyl Ether	10.0
Perfume	5.0

Blend all ingredients at room temperature.

No. 3

(Dispersible)

Drakeol 7 (USP)	60.0
Isopropyl Myristate	25.0
Liquid Lanolin	10.0
PEG 40 Sorbitan Peroleate	5.0
Perfume	q.s.

Blend all ingredients at room temperature.

No. 4

Mineral Oil	70
Olepal® Isostearique	10
Labrafil® Isostearique	10
Oily Vegetol® Marigold	5
Perfume	5

No. 5

(Floating)

Mineral Oil	46
Propylene Glycol Dipelargonate	48
Hydrophilol® ISO	1
Perfume	5

At room temperature, mix the components, and add perfume.

No. 6

(Foaming)

A	Sodium Xylene Sulfonate	21.000
	Sipon® LT6	7.000
	Superamide L9	5.600

B	Propylene Glycol	1.400
	Sodium Chloride	0.350
	Formalin	0.035
	SDA-40 Alcohol	0.350
	Deionized Water	33.920
	Lantrol® AWS	0.350
C	**Trisolan**	3.000
	Lantrol®	6.000
	Tenox 2	0.030
	Peach Kernel Oil	19.470
	Perfume	1.500

Heat Part A until it is a uniform solution. Add Part A to Part B. Adjust pH to 6.3 with dilute lactic acid. Prepare Part C. Mix together until clear and package.

Blooming Bath Lotion

A	**Cutina GMS**	4.00
	Eumulgin® B-1	4.00
	Cetiol G-20S	15.00
	Cetiol LC	15.00
B	Water	60.50
	Propylene Glycol	1.50
C	Fragrance, Preservative	q.s.

Mix and heat Part A at 60–65 C. Mix and heat Part B at 60–65 C and add to Part A. Mix until cooled to 40 C and add Part C.

After-Bath Lotion With Talc

Formula No. 1

A	Deionized Water	45.84
	Carbopol® 941 (2% sol'n.)	15.00
	Propylene Glycol	5.00

B	**Cerasynt 945**	4.00
	Ceraphyl 140	12.00
	Cetyl Alcohol	1.00
	Escalol® 507	3.00
	Siloxane SWS-03314	5.00
	Promulgen® D	2.00
	Germaben® II	1.00
C	**Talc 141**	5.00
D	Sodium Hydroxide (10% sol'n.)	1.16

Heat Part A and Part B to 80 C. Add Part B to Part A. Begin cooling and slowly add Part C. Mix until talc is well dispersed. At 60 C, add Part D. Cool to room temperature with continuous mixing.

No. 2

A	**Cerasynt 945**	4.00
	Ceraphyl 140-A	12.00
	Cetyl Alcohol	1.00
	Escalol® 507	1.50
	Siloxane SWS-03314	5.00
	Promulgen® D	2.00
B	Deionized Water	47.34
	Carbopol® 941 (2% sol'n.)	15.00
	Propylene Glycol	5.00
C	**Talc 141**	5.00
D	**Germaben® II**	1.00
E	Sodium Hydroxide (10% sol'n.)	1.16

Heat Part A and Part B to 80 C, add Part A to Part B and begin cooling. Add talc slowly, mixing well to fully disperse. Below 60 C add Part D and Part E. Cool to room temperature.

After-Bath Body Splash

(Alcoholic)

SDA-40 Alcohol	75
Isopropyl Myristate	15
Ross Jojoba Oil	2
Fragrance	8

Mix all ingredients in a stainless steel vessel, run through appropriate filter, and package.

After-Bath Gel

A	Demineralized Water	86.55
	Kathon CG	0.10
B	Carbomer 940	0.50
C	**Vegetol® White Lily** (water soluble)	5.00
	Vegetol® Mallow (water soluble)	5.00
	Perfume	0.20
	Solvating Agent PL 12 γ 2428	0.40
	Sodium Hydroxide (10% aq. sol'n.)	2.25

Disperse Carbomer in Part A under fast stirring. Then other components are added while slow stirring. Avoid the introduction of air. Add sodium hydroxide sol'n.

Eyeliner

Formula No. 1

(Waterproof)

Beeswax	16.5
Ganex V-220	5.0
Shell Sol® 71	35.0
Bentone® Gel SS71	33.5
Preservative	0.2
Chroma-Lite® Black	9.8

No. 2

(Liquid)

Water	80.75
Sodium Carboxymethyl Cellulose	1.50
Hectorite	0.50
Propylene Glycol	5.00
Triethanolamine-Shellac (25%)	8.00
Iron Oxide	4.00
Methyl Paraben	0.10
Germall® II	0.15

Eye Shadow Base

A	Stearic Acid (triple-pressed)	35.0
	Penreco Super (USP)	6.0
	Lanolin	5.0
B	Propylene Glycol	5.0
	Triethanolamine	4.0
	Water	45.0

Heat Part A to 75 C and Part B to 80 C. Add Part B to A. Continue to stir until the mixture cools. Appropriate pigments may be added to arrive at the desired color.

Note: This formula produces a good base for cream eye shadow. Creamy, smooth, and moist, the base readily accepts pigments.

Eye Shadow Stick

Drakeol 7 (USP)	41.0
Ozokerite	23.8
White Beeswax	15.9
Butyl Stearate	2.9
Candelilla Wax	7.2
Cetyl Alcohol	9.2

Heat the ingredients to 85 C, mix and add the desired amounts of color. Cool with agitation to 70 C and mold.

Pressed Powder Pearlescent Eye Shadow

A	**Veegum® F**	5
	Timica Pearl White	35
	Talc	29
	Zinc Stearate	8
	Magnesium Carbonate	1
B	**Acetol® 1706**	3
	Tween® 20	9
	Water	10
	Preservative	q.s.

Micropulverize Part A. Mix Part B until uniform and add to Part A. Mix until uniform. Screen through a No. 16 sieve and compress to desired hardness.

Consistency: Firm cake.

Suggested Packaging: Compact.

Note: **Veegum® F** serves as the binder in this formula giving a firm cake. To apply to eyelid, wet brush and rub on cake surface.

Eye Shadow Pomade

Syncrowax ERL-C	6.0
Syncrowax HR-C	2.8
Propylene Glycol Dicaproate/Caprylate	10.0
Volatile Silicone 7158	47.2
Talc (USP)	14.0
Color	10.0
Timica Sparkle	10.0

Mill the colors in a combination of **Volatile Silicone** and **Crodamol PC** to make a paste. Melt waxes and add rest of silicone. Disperse **Timica** and talc under low shear in oils. Add color blend. Place in 60 C water bath to deaerate. Fill off into pots at 45–50 C.

Creamy Eye Shadow

		Formula No. 1	No. 2
A	**Ceraphyl 28**	2.00	2.00
	Ceraphyl 847	2.00	2.00
	White Beeswax	7.80	7.80
	Ganex V-220	6.50	6.50
	Thixcin R	3.50	3.50
	Shell Sol® 71	33.00	28.00
B	**Bentone® Gel SS-71**	14.70	19.70
	Zinc Stearate	2.00	2.00
	Magnesium Stearate	1.00	1.00
	Titanium Dioxide 328	6.00	2.00
	Talc 141	14.50	12.50
	Chroma-Lite® Light Blue	6.00	—
	Chroma-Lite® Green	—	12.00
C	**Germaben® II**	1.00	1.00

Combine and mix Part A with propeller stirrer while heating to 75 C or until melted. Add Part B ingredients and mix with homo-mixer at medium speed. Stir with propeller stirrer while cooling. Mix Part C at 60 C. Cool to just above congealing point and pack.

Mascara

Formula No. 1

(Solvent)

A	**Emerwax 1253**	10.00
	Ross Ozokerite Wax 170	10.00
	Cetyl Alcohol N.F.	8.00
	Yellow Carnauba Wax (#1, flakes)	2.00
	Lanfrax	0.50
B	Mineral Spirits	59.50
C	Pigments	10.00

Dissolve all the waxes in the mineral spirits at 90 C. Employ a reflux column. Allow the mixture to cool to 50 C. Add pigments under high-speed agitation. Grind through a heated three roller, ointment or colloid mill. Add aluminum stearate at 1–3 percent as a gellant for viscosity control if necessary.

No. 2

(Waterproof)

A	**Shell Sol® 71**	40.60
	Bentone® Gel SS-71	15.00
B	**Talc 141**	1.00
	C33-5198 Cosmetic Black J	10.00
C	**Ceraphyl 85**	2.00
	Ceraphyl 375	2.00
	White Beeswax	10.00
	Ozokerite	10.00
	Thixcin R	3.00
	Ganex V-220	6.00
	Methylparaben	0.10
	Butylparaben	0.10
	Sorbic Acid	0.20

Admix Part B and pulverize until uniform. Mix Part A with homomixing agitation. Add Part B slowly to Part A while maintaining homomixing agitation, then heat to 70 C. Mix and heat Part C to 75 C, then combine with Parts A and B. Cool to just above congealing point and fill into containers.

Eye Makeup Remover

Miranol® 2MHT Modified	5.0–10.0
Tween® 20	1.0–2.0
Propylene Glycol	1.0–2.0
Mixed Parabens	0.1–0.1
EDTA	0.1–0.1
Water	92.8–85.8

Mix the ingredients together and heat to 50–55 C. Adjust the pH to 7.0–8.0.

Physical Properties:
Solids — 4.6–9.1%

Reticulin Eye Cream

A	**Polawax®**	4.0
	Stearic Acid (triple-pressed)	3.0
	Crodamol PMP	14.0
	Crodacol C-95	2.0
	Super Corona Lanolin	5.0
	Syncrowax HGL-C	3.0
	Propyl Paraben	0.1
	Propylene Glycol	5.0
	Veegum®	1.5
	CMC 7MF	0.5
B	Methyl Paraben	0.2
	Germall® 115	0.3
	Deionized Water	58.4
	Triethanolamine (99%)	1.0
	Reticusol	1.0
C	**Crolastin**	1.0

Combine Part A and heat to 75–80 C. Combine Part B with good agitation and heat to 75–80 C. Slowly add Part A to B maintaining temperature and mixing. Allow to mix 10 min at 75 C, then cool at 50 C and add Part C. Continue cooling.

Note: This system can be used prior to makeup application. A fragrance suitable for use in sensitive skin areas is **Hedione** at levels up to 2%.

Rouge

Formula No. 1

(Cream)

A	**Ross Refined #1 Yellow Carnauba Wax**	6
	Ross Ozokerite Wax 77W	10
	Mineral Oil	24
	Isopropyl Palmitate	27
B	Talc	10
	Titanium Dioxide	20
	Color	3

Melt Part A to 70 C. When cooled run together with Part B on a 3-roll mill.

No. 2

(Solid)

Isopropyl Myristate	45.0
Drakeol 7 (USP)	22.5
Pigment, as desired	19.7
Candelilla Wax	8.8
Lanolin	4.0

Mix the ingredients together and heat to 85 C with continuous stirring. Cool to 70 C and mold.

Gel Blusher

Base:

Carbopol® 934	1.0000
Water	76.4000
Sodium EDTA	0.0001
Glycerin	16.0000
Propylene Glycol	5.0000
Methyl Paraben	0.20000
Propyl Paraben	0.20000
Triethanolamine	to pH 7.2

Formula:

Base	85.0
Chroma-Lite® Magenta	2.0
Bi-Lite® 20	13.0

Slowly add **Carbopol®** to water; stir until clear. Continue to add ingredients progressively down the list slowly.

Note: This is an excellent vehicle for cosmetic colors.

Makeup Base

Formula No. 1

(Multipurpose)

Syncrowax HGL-C	12.0
Syncrowax HR-C	2.0
Syncrowax HRS-C	1.5
Klearol	49.7
Fluilan	5.0
PVP K30	0.8
Timica Sparkle	20.0
Talc (USP)	1.5
Color	7.5

No. 2

(Liquid, All-Purpose)

Oil Phase:

Amerchol L-101®	6.0
Acetulan®	4.0
Glycerol Monostearate (neutral)	2.0
Stearic Acid	5.0

Water Phase:

Solulan® 98	2.0
Propylene Glycol	5.0
Triethanolamine	1.0
Color Blend	5.0

Water	67.5
Pearl-Glo® M	2.0
Perfume and Preservative	q.s.

Add the water phase at 85 C to the oil phase at 85 C with mechanical agitation. Cool with stirring to 40 C. Add perfume and continue cooling with stirring to room temperature.

No. 3

Oil Phase:	
Amerchol L-101®	6.0
Acetulan®	4.0
Glycerol Monostearate (neutral)	2.5
Stearic Acid (triple-pressed)	5.0
Water Phase:	
Solulan® 98	2.0
Propylene Glycol	5.0
Triethanolamine	1.0
Water	74.5
Fragrance	q.s.

Add water phase at 85 C to the oil phase at 85 C with stirring. Cool to 40 C with stirring and add perfume, continue cooling to room temperature with stirring.

Liquid Makeup

Formula No. 1

(Pan)

Syncrowax HGL-C	12.00
Syncrowax HRC	3.00
Super Refined Sesame Oil	7.00
Crodamol PMP	44.25
Stearic Acid (triple-pressed)	3.00
Talc 1003	13.00

Brown Oxide A1100	0.75
Yellow Oxide A1301	1.50
Red Oxide A1249	0.75
Titanium Dioxide	14.00
Cab-O-Sil® M5	0.75

Combine powders in blend or micro pulverizer and pulverize until no pigment streaks appear on draw down. Combine oils and waxes in batch vessel and heat to 85 C with agitation. Ensure all solids are completely melted. Maintain temperature and agitation and slowly add powder blend to oils. Allow to mix at 85 C for 15 min or until powders are uniformly dispersed. Cool to 55 C and fill into compacts.

No. 2

(Low Comedogenic)

Propylene Glycol (USP)	6.00
Triethanolamine	0.95
Methyl Paraben	0.20
CMC 12 M 31P	0.60
Veegum® HV	0.80
Deionized Water	64.30
Titanium Dioxide #3328	6.00
Talc #5141	1.00
Cosmetic Pink #7153	1.00
Cosmetic Yellow #7156	2.00
Crodamol PMP	11.00
Crodafos N10 Neutral	2.00
Polawax®	2.00
Syncrowax AW1-C	2.00
Propyl Paraben	0.15

Disperse the **Veegum®** in 25 percent of the deionized water. Separately combine the propylene glycol and triethanolamine and mix. In this mixture dissolve the methyl paraben, disperse the CMC and then add the remainder of the water, which has been previously heated to 85 C, to hydrate the gum. Mix well, and when a clear solution is obtained, add the **Veegum®** dispersion. With rapid agitation, slowly add the mixture of pigments, talc, and titanium dioxide which has been thoroughly pulverized in a mortar. Hold temperature at 80 C and when the solids are thoroughly dispersed, add the previously heated (80 C) mixture of oils and waxes while stirring vigorously. Hold at 80 C for 5 min and allow to cool. At 50 C replace water lost through evaporation and continue mixing and cooling until temperature decreases to 35 C.

Cream Matte Makeup

A	**Veegum®**	2.60
	CMC 7LF	0.40
	Water	42.70
B	Propylene Glycol	5.00
	Water	12.30
C	Talc	18.50
	Vanclay Kaolin	1.30
	Titanium Dioxide	3.70
	Iron Oxides	1.50
D	Isopropyl Myristate	5.00
	Arlacel® 20	0.75
	Tween® 20	2.25
	Stearyl Alcohol	2.00
	Amerchol L-101®	2.00
	Preservative	q.s.

Blend **Veegum®** and CMC. Slowly add to the water, while agitating at maximum available shear. Continue mixing until smooth. Micropulverize Part C. Add to Part B and grind to a smooth paste. Add to Part A and heat to 65 C. Heat Part D to 70 C and add to Parts A/B/C. Mix until cool.

Consistency: Medium viscosity cream.

Suggested Packaging: Jar or tube.

Note: **Veegum®** stabilizes the emulsion and provides uniform pigment suspension. This nonionic makeup spreads smoothly and evenly with a light greaseless feel. It can be applied with a wet sponge if desired. With slight modifications, this formula can be used as a cream eye shadow.

Waterproof Makeup

Oil Phase:	
Witconol 14	3.00
Aluminum Stearate	1.50
Aristowax 143 (143 F m.p.)	5.00
Mineral Oil (70 SUS visc.)	23.50
Pigment Blend (see below)	8.00
Water Phase:	
Germall® II	0.20
Methyl Paraben	0.15
Propyl Paraben	0.10
Deionized Water	58.55
Pigment Blend:	
Titanium Dioxide (oil dispersible)	87.80
Low Micron Umber	5.20
Cosmetic Yellow Iron Oxide	5.20
Cosmetic Red Iron Oxide	1.80

Heat oil phase to 90–95 C and stir 10–15 min. Disperse premixed pigment blend in hot oil phase and stir for 15 min. Heat water phase to 90–95 C. When preservatives are dissolved, add water phase (at 90–95 C) to oil phase with moderate agitation while maintaining temperature at 90–95 C. When all the water phase has been added continue stirring and maintain temperature for additional 15–20 min. Then let cool with stirring and package at 28 C in containers. Cooling period is 3 h.

Makeup Mousse

		Formula No. 1	No. 2
A	**Emulsynt GDL**	2.00	2.00
	Ceraphyl 840	1.00	1.00
	Stearic Acid (triple-pressed)	5.00	5.00
	Cetyl Alcohol	0.50	0.50
	Escalol® 507	2.00	2.00
	Ceraphyl 140-A	4.00	10.00

B	Deionized Water	49.90	43.90
	Methocel® K4M	0.10	0.10
	Propylene Glycol	2.00	2.00
	Pigment Blend #F96-7 (see below)	12.00	12.00
C	Triethanolamine (88%)	0.50	0.50
D	SD-40 Alcohol	20.00	20.00
	Germaben® II	1.00	1.00

Disperse **Methocel®** in water and heat to 80 C. Add propylene glycol to water. Slowly add Pigment Blend to water phase, mixing until all of the powder is well dispersed. Heat Part A to 80 C. Add Part A to Part B, mixing well. Add triethanolamine to batch, mixing well to fully neutralize. Cool batch to 35 C and add Part D. Cool to room temperature and fill cans.

Fill:	Concentrate	95
	Propellant A-46	5
	Precision Valve 2X.020 inverted	
	Foam Spout #02-1512	

Pigment Blend #F96-7:

Titanium Dioxide #3328	82.50
Cosmetic Brown Iron Oxide C33-5136	0.50
Cosmetic Yellow Iron Oxide C33-8073	14.00
Cosmetic Iron Oxide Red 7051	3.00

Makeup Remover

Distilled Water	53.8
Alconate CPA	20.0
Sipex® EC-111	16.0
Sipoteric 1398	10.0
Glydant®	0.2
Fragrance, Color	q.s.

Add **Alconate** to water while heating to 65 C with moderate agitation. Add **Sipex** and agitate until a uniform dispersion results. Discontinue heat.

Add **Sipoteric** and stir to a uniform dispersion. Add **Glydant** and cool to below 50 C for fragrance addition.

Physical Properties:
Appearance — Opaque liquid
pH — 7.7
Visc. (#3 spindle @ 6 rpm) — 1800 cps

Lipstick

Formula No.1

A	Castor Oil	3.3
	Lanalane	54.7
	Lipovol A	10.0
	Myristyl Lactate	5.0
	Carnation Mineral Oil	5.0
B	Carnauba Wax	8.0
	Durmont 500 D	7.0
	Castor Oil (in the color blends)	6.7
C	Perfume	0.3
D	Color Blends	14.6

Heat the oils of Part A to 95 C in a steam-jacketed kettle. Add the waxes of Part B and maintain the temperature at 95 C. Heat the waxes and oils to 110 C while mixing and hold for 15–20 min or until all the waxes are in solution. Cool the batch to 95 C and transfer approximately 15 parts to a small pot and maintain heat at 90–95 C. Add the color blend to the base in the small pot slowly, stirring vigorously. Continue mixing until all the color is completely dispersed. Add the perfume from Part C to the color in the small pot. Mix until completely dispersed. Transfer this color dispersion to the balance of the base which is at 95 C in the steam-jacketed kettle. The base should be stirred while the color is being added. Mix until completely dispersed.

No. 2

A	Castor Oil	3.3
	Lanalane	19.7
	Lipovol A	15.0
	Amerlate® P	30.0
	Lanolin (anhyd.)	10.0
	Castor Oil (in the color blends)	6.7
B	Candelilla Wax	5.0
	Durmont 500 D	5.0
	Vestowax FT 150	5.0
C	**A80-264 Perfume**	0.3
D	Color Blends	14.6

Heat the oils of Part A to 95 C in a steam-jacketed kettle. Add the waxes of Part B and maintain the temperature at 95 C. Heat the waxes and oils to 110 C while mixing and hold for 15–20 min or until all the waxes are in solution. Cool the batch to 95 C and transfer approximately 15 parts to a small pot and maintain heat at 90–95 C. Add the color blend to the base in the small pot slowly, stirring vigorously. Continue mixing until all the color is completely dispersed. Add the perfume from Part C to the color in the small pot. Mix until completely dispersed. Transfer this color dispersion to the balance of the base which is at 95 C in the steam-jacketed kettle. The base should be stirred while the color is being added. Mix until completely dispersed.

No. 3

(High Shine)

Castor Oil	59.4
Candelilla Wax	8.0
Acetulan®	7.5
Rosswax 1275W	5.0
Propylene Glycol Monolaurate	5.0
Lanogene	5.0
Carnauba Wax	2.0
Propyl Paraben	0.1

Timicon MP-10	7.0
D & C Red #9	0.6
D & C Red #7 Ca Lake	0.3
Pur. Navy Blue #7110	0.1
Fragrance	q.s.

Grind pigments in part of castor oil using either a 3-roller mill or mortar/pestle. Add all other ingredients (except for pearlescent pigment and fragrance) and heat gently on steam bath to 80–85 C. Add pearl, mix until homogeneous. Fragrance should be added at lowest possible temperature. Cast into molds.

No. 4

(Moisturizing)

A	**Ross White Ozokerite Wax #1544**	3.00
	Ross Refined Candelilla Wax	8.00
	Ross Refined #1 Yellow Carnauba Wax	3.00
	Ross White Beeswax	2.00
	Ceraphyl 50-S	5.00
	Ceraphyl 375	7.60
	Crystal Crown	30.40
	Penreco 7	10.00
	Lantrol® 1674	5.00
	Propyl Paraben	0.15
	Antioxidant	0.05
B	**#6905 D&C Orange #5-Zr. Lake**	4.50
	Ceraphyl 847	10.50
C	**#6627 D&C Red #27-Zr. Lake**	0.76
	Ceraphyl 847	1.14
D	FD&C Blue # A1. Lake in Castor Oil	1.60
E	**Lustra-Pearl Gloss**	5.50
	Lustra-Pearl Gold	1.50
F	Fragrance	0.30

Prepare Part B and Part C using two passes over a three-roll mill at room temperature. Combine Part A and heat to 85 C while stirring until melted clear and homogeneous. Cool to 75 C and add Parts B, C, D, and E to it and after each addition mix thoroughly. Cool to just above congealing point. Mix Part F and pour into molds.

Lip Gloss

Formula No. 1

Drakeol 7 (USP)	45.0
Penreco Super White Petrolatum USP	20.0
White Beeswax	25.0
Isopropyl Myristate	10.0

Mix the ingredients together and heat them to 85 C under anhydrous conditions. Continue stirring and maintain temperature until the beeswax is entirely melted. Cool the mixture to 70 C with stirring and mold.

No. 2

(Lustrous)

A	Carnauba	2.00
	Ozokerite	2.00
	Lanfrax 1777	5.00
	Indopol® H-100	40.00
	Lantrol® 1674	36.70
	Ceraphyl 50-S	3.00
	Ceraphyl 847	7.00
	Propyl Paraben	0.10
	Antioxidant	0.10
B	**Spectra-Pearl® RDW**	0.80
	Lustra-Pearl Glimmer	3.00
C	**Fragrance D84-389**	0.30

Combine Part A and heat to 80–85 C until melted and homogeneous. Cool to 75 C, add Part B ingredients in the order written and mix thoroughly.

At 55 C mix Part C and cool to just above congealing point and package.

Sunscreen Lip Balm

Formula No. 1

Super Sterol Ester	10.00
Super Refined Apricot Kernel Oil	53.75
Syncrowax HGL-C	12.00
Syncrowax ERL-C	6.00
Silicone Copolymer F755	1.00
Fluilan	5.00
Escalol® 507	7.00
Uvinul® N539	3.00
Propyl Paraben	0.20
Veragel Lipoid 1:1	1.00
BHA	0.05
Fragrance 14390	1.00

Combine ingredients and heat to clarity, with agitation, approximately 80 C. Cool. Pour at 54–58 C.

No. 2

Stick Base CB 8829	79
Titanium Dioxide	5
Propylene Glycol Dipelargonate	15
Perfume	1

Heat base to 80 C, then add the other components of formula. At about 75 C, pour the preparation in the sticks mold.

Antichap Stick

	Formula No. 1	No. 2	No. 3
A **Protopet Alba**	65	65.0	65.00
Standamul® 1616	12	12.0	12.00
Crodacol C95 NF	5	5.0	5.00

B	**Vestowax FT 150**	3	3.0	3.00
	Duron Ozokerite 71/74	15	15.0	15.00
C	Color	—	q.s.	q.s.
D	**Tegosept M**	—	—	0.20
	Tegosept P	—	—	0.10
	BHA	—	—	0.02
E	**Coviox T-70**	—	0.5	—
	Allantoin	—	0.1	—
F	Perfume*	—	q.s.	q.s.

Heat the ingredients of Part A to 95 C in a steam-jacketed kettle. Add the waxes of Part B. Heat the waxes and oils to 110 C while mixing and hold for 15–20 min or until the waxes are in solution. Cool the batch to 95 C and transfer approximately 15 parts to a small pot and maintain the heat at 90–95 C. Add the color of Part C to the base in the small pot slowly while stirring vigorously. Continue mixing until all the color is completely dispersed. When all the ingredients have been added to the base material in the pot and completely dispersed, transfer the materials in the pot to the balance of the base in the kettle. (*Note:* The material from the pot should be added slowly to the base in the kettle. The material in the kettle should have good top-to-bottom mixing using a lightnin mixer.) Mixing should continue until the entire batch is homogeneous. The batch can be filled or stored. If stored, polyethylene-lined cartons should be used. A 10-kilo capacity is satisfactory. When remelting this material the entire contents of the carton should be used as separation may have occurred.

Note: This formulation has been designed to develop the necessary hardness for good heat resistance without exhibiting the flaking characteristics common to some commercial products. It will provide excellent occlusiveness and smooth application glide with no tack.

---

* Subtract from **Protopet Alba**. 100 is total after designated subtractions.

Lip Conditioning Cream

Polawax®	10.00
Super Refined Soy Bean Oil	5.00
Super Refined Shark Liver Oil	5.00
Silicone 101 (350 cs)	0.50
Indopol® H1900	2.00
Paraffin Wax 140	3.00
BHA	0.05
Deionized Water	62.45
Glycerin USP	5.00
Collasol	2.00
Crolastin	2.00
Reticusol	2.00
Germaben® II	1.00

Combine water phase and heat to 78 ± 2 C. Combine oil phase and heat to 78 ± 2 C under controlled agitation. Slowly add water phase to oil phase. Cool. At 45 C add protein phase and preservatives. Cool to room temperature and fill off.

Cuticle Massage Cream

Oil Phase:

Super Refined Apricot Kernel Oil	5.00
Carnation Mineral Oil	5.00
Super Hartolan	5.00
Polawax®	10.0
Crodacol C-95	3.0
Glyceryl Monostearate (pure)	2.0

Water Phase:

Deionized Water	62.0
Glycerin	5.0
Germaben® II	1.0
Crotein K	2.0

Combine oil phase and heat to 78 ± 2 C. Combine water phase and heat to 78 ± 2 C. Add water phase to oil phase under agitation. Cool. At 40 C add **Crotein.** Cool to room temperature.

Nail Lacquer

A	Isopropanol	2.0
	Butyl Acetae	33.5
	Ethanol	3.0
	Toluene	25.5
B	Nitrocellulose	15.0
C	**Santolite**	7.0
D	Dioctyl Phthalate	4.0
E	**Bentone® 34**	2.0
F	**Bi-Lite® 20**	4.0
	Color Blend	6.0

Measure 3/4 of each ingredient of Part A. Pour this into a container that is fitted with a high-shear stirrer. Then add Part B. Stir until dissolved. Then add Part C and stir until dissolved. Add Part D and stir until completely mixed. Slowly add Part E and the remaining portion of Part A. Allow to stir at a high rate of shear. With a lower rate of shear add the color blend and **Bi-Lite 20** pigment. Stir until uniform.

Nail Hardner/Strengthener

Nitrocellulose	10–15
Isopropyl Alcohol	5–7
Polyester Resin	4–6
Acrylates Copolymer	4–6
Toluene Sulfonamide/Formaldehyde Resin	3–5
Dibutyl Phthalate	5–7
Butyl Acetate	30–35

Face Mask

Formula No. 1

(Peelable)

A	**Veegum®**	1.5
	Water	58.5
	Vinol® 523	10.0
	Propylene Glycol	7.0
B	SD Alcohol 40	20.0
	Hetoxol OL-23	3.0
	Preservative	q.s.

Slowly add **Veegum®** to the water, while agitating at maximum available shear. Continue mixing until smooth. Add propylene glycol with stirring and heat to 70–75 C. Slowly add **Vinol® 523**, mixing until smooth and uniform. Continue mixing until 40 C. Dissolve **Hetoxol OL-23** in the alcohol. Add Part B to A and mix until smooth and uniform.

Consistency: Fluid lotion.

Suggested Packaging: Wide mouth jar with applicator brush.

Note: **Veegum®** thickens this formula and allows smooth application without tackiness or stringiness. The mask should be applied with a brush, spreading a thin uniform film on the skin. After approximately 10 min, rinse or peel off.

No. 2

(Defectiveness Skin)

A	**Tefose® 2561**	13
	Oil Vegetol® Marigold WL 1072	4
B	Demineralized Water	62
	Fondix® G BIS	1
	Bentonite	5
	Zinc Oxide	10
C	**Vegetol® T**	5

Part A and Part B are heated to 75 C. Part B is added to Part A. Cool while stirring slowly. Add Part C. Homogenize if necessary.

No. 3

(Seaweed)

A	Demineralized Water	41.9
	Kathon CG	0.1
	Kaolin	5.0
	Bentonite	3.0
	Calcium Carbonate	25.0
	Tefose® 1500	10.0
	Mineral Oil	3.0
	Fondix® G BIS	1.0
	Glycerin	5.0
B	**Vegetol® Seaweed** (water soluble)	5.0
	Perfume	1.0

Part A is heated to 75–80 C. Cool while stirring slowly. At about 35 C, add other components.

No. 4

Gaffix™ VC-713	10.0
Ethanol	33.9
Dow Corning® 190	0.1
Distilled Water	50.0
Gelvatol 20/90	6.0
Color, Fragrance, Preservative	q.s.

Step 1 Combine first three ingredients listed, stirring until dissolved.
Step 2 Combine and dissolve the **Gelvatol** and water.
Step 3 Add Step 1 to Step 2. Stir until uniform.

Beauty Mask

Formula No.1

Sodium Alginate	3.07 g
Magnesium Oxide (USP heavy)	4.11 g
Magnesium Carbonate (NF powder)	3.07 g

Calcium Sulfate (NF powder)	2.56 g
Tetrasodium Pyrophosphate	0.82 g
Perfume, Active Ingredients	0.07 g
Water	50.00 g
Bi-Lite® 20	3.07 g

The above dry mix should be well blended, preferably in a ball mill. 13.7 g of the mixture is stirred for about 1 min with 50 g of water. The homogeneous paste is applied to the skin, where it sets to a gel-like film which can be removed after 15 min.

No. 2

A	**Vinol® 523**	10.00
	Propylene Glycol	5.00
	Water	44.50
	Carbopol® 961	0.400
	Propyl Paraben	0.025
B	Ethyl Alcohol	30.000
	Methyl Paraben	0.050
	WSP-X250	5.000
	Color Blend	0.025
	Chroma-Lite®	5.000

Physical Properties:
pH—7.2

Heat water and propylene glycol to 70–75 C. Add **Vinol® 523** with stirring. Stir to dissolution. Add **Carbopol® 961** with stirring at 70–75 C until it dissolves. In a separate vessel, combine Part B (no heat required). Cool batch to 40 C. Add Part B slowly while stirring. Add **WSP-X250** at 40 C. Cool batch to 30 C and fill.

Note: Both Formulas No. 1 and No. 2 are easily peelable facial masks which are noticeably effective.

Tightening Cream Base

A	**Lexemul® 561**	6.00
	Lexemul® EGMS	2.00
	Lexol 3975	5.00
	Lexate PX	5.00
	Lexate IL	2.00
	A-C® Polyethylene 617A	5.00
	Mineral Oil	43.00
	Lexgard P	0.05
B	Sodium Lauryl Sulfate (powder)	0.50
	Lexgard M	0.15
	Water	q.s. to 100

Melt contents of Part A together. Heat water to 80 C and add SLS and **Lexgard M**. Slowly add Part B to Part A with thorough mixing or homogenization until the set point is reached.

Note: This emulsion seems to tighten the skin. It can be used as a medicinal cream base or adapted for use as a temporary wrinkle remover.

Physical Properties:

pH — 6.5
Visc. (@ 24 C) — 20,000 cps
Stability after 1 month — Stable at room temperature at 45 C and passes 3 freeze/thaw cycles.

Wrinkle Smoother

A	**Veegum®**	1.5
	CMC 7LF	1.0
	Water	82.5
B	**Flexan® 130**	12.0
C	**Collasol**	3.0
	Preservative	q.s.

Blend **Veegum®** and **CMC**. Slowly add to the water, while agitating at maximum available shear. Continue mixing until smooth. Add Parts B and C in order, mixing after each addition until smooth and uniform.

Consistency: Fluid lotion.

Suggested Packaging: bottle.

Facial Moisture Lotion

Formula No. 1

Incroquat BTQ	1.00
Crodacol CS-50	0.50
Super Sterol Ester	0.50
Crodamol PMP	1.00
Polawax®	1.50
Crosilk Liquid	5.00
Volatile Silicone 03314	0.50
Deionized Water	89.00
Germaben® II	1.00

Combine oil phase and heat to 80 C with controlled agitation. Combine water phase with protein phase and heat to 82 C. Slowly add water phase to oil phase. Cool. At 50 C add **Volatile Silicone.** Cool to room temperature.

No. 2

(With Sunscreen)

Incroquat Behenyl TMS	1.00
Crodacol CS-50	0.50
Super Sterol Ester	0.50
Crodamol PMP	1.00
Polawax®	3.00
Parsol 1789	1.50
Parsol MCX	2.00
Citric Acid	0.10
Crotein CAA SF	5.00
Volatile Silicone 03314	0.50
Deionized Water	84.40
Hampene Na_2	0.20

Methyl Paraben	0.20
Propyl Paraben	0.10

Combine oil phase and heat to 80 C with controlled agitation. Combine water phase with protein phase and heat to 82 C. Slowly add water phase to oil phase. Cool. At 50 C add **Volatile Silicone**. Cool to room temperature.

Hand Lotion

Formula No. 1

(Cationic)

Oil Phase:	
Amerchol L-101®®	5.0
Solulan® 98	2.0
Cetal	1.0
Glyceryl Stearate (and) PEG-100 Stearate	4.0
Water Phase:	
A **Quatrisoft™ Polymer LM-200**	0.5
Water	19.5
B Steapyrium Chloride	0.1
Glycerin	2.0
Water	65.9
Preservative and Perfume	q.s.

Add **Quatrisoft™ Polymer LM-200** to the water (water phase, Part A). Mix until thoroughly dispersed, then heat to 45–50 C. Mix until clear and uniform. Add Water Phase Part A to Water Phase Part B with mixing while heating to 70 C. Heat Oil Phase to 70 C. Add Water Phase at 70 C to Oil Phase at 70 C while mixing. Cool to 35 C.

No. 2

(Glycerin)

A **Arlasolve 200**	0.20
Fragrance H/45755	0.04

B	Water	53.91
C	**Natrosol® 250HR**	0.85
	Glycerin	25.00
D	**Arlatone PQ 220**	20.00
	Germaben® II	1.00

Disperse **Natrosol®** in glycerin. Mix Part A with gentle heat until uniform. Add water slowly to Part A with agitation. When uniform, add Parts A and B to Part C. Apply heat and stir until a uniform gel develops. Add Part D and stir until uniform.

Physical Properties:

Typical visc. — 3100 cps
Stability — 1 month at 25 C, 40 C, and 50 C

No. 3

(Jojoba)

A	**Modulan®**	1.6
	Amerchol L-101®	0.8
	Isopropyl Palmitate	5.0
	Glyceryl Monostearate Pure	2.1
	Rosswax 63-0412	4.0
	Isopropyl Myristate	4.0
	Ross Jojoba Oil	1.6
B	Water	74.4
	Emery® 916	4.2
	Triethanolamine	2.3

Melt Parts A and B in separate vessels to 170 F under agitation. When temperature is reached, mix Part A into Part B and cool. Package containers at below 120 F.

No. 4

A	**Lexemul® LPM**	2.50
	Lexemul® 561	2.00
	Lexol IPP	3.00
	Stearic Acid	1.00
	Lexgard P	1.00
B	Propylene Glycol	1.00
	Carbomer 934	0.20
	Triethanolamine	q.s. to pH 7.0
	Lexgard M	0.15
	Water	q.s. to 100

Heat water to 70 C. Disperse Carbomer 934 and neutralize. Dissolve **Lexgard M** in propylene glycol and add to Part B. Separately melt contents of Part A together. Slowly add Part A to Part B with agitation. Mix and cool until the set point is reached, then fill.

Physical Properties:

pH — 7.0
Visc. (@ 24 C)— 18,000 cps
Stability after one month — Stable at room temperature at 45 C, and passes 3 freeze/thaw cycles

Body Lotion

Formula No. 1

A	**Tefose® 2561**	7.0
	Isostearyl Stearate	5.0
B	Demineralized Water	50.0
C	Demineralized Water	27.8
	Kathon CG	0.1
	Phylderm® Filatov	10.0
D	Perfume	0.1

Parts A and B are heated to 75 C. Part B is added to Part A. Cool while stirring slowly. At about 25 C, add Part C. Add Part D as required. Homogenize if necessary.

No. 2

A	**Tefose® 2561**	6.00
	Mineral Oil	6.00
	Lanolin Oil	2.00
B	**Kathon CG**	0.10
	Demineralized Water	45.45
	Carbopol® 934	0.10
C	Perfume	0.20
D	Demineralized Water	30.00
	Triethanolamine	0.15
	Amniotic Fluid	10.00

Parts A and B are heated to 75 C. Part B is added to Part A. Cool while stirring slowly to 25 C. Add Part C then add Part D. Homogenize if necessary.

All-Purpose Lotion

Formula No. 1

(Nonoily)

A	**Cutina E-24**	2.00
	Cutina MD-A	6.00
	Lanette O	1.50
	Cetiol LC	8.00
B	Glycerin	5.00
	Deionized Water	75.30
C	**Kathon CG**	0.10
	Fragrance	0.10

Heat Part A to 70–75 C. Heat Part B to 70–75 C. While agitating add Part B to Part A. Remove heat and continue mixing during cooling step. When the batch temperature has reached 40–45 C, add the components of Part C. Continue mixing for 15–30 min. Fill off.

No. 2

(Conditioning)

Oil Phase:	
Promulgen® G	5.0
Glucate® SS	2.0
Glucamate® SSE-20	2.0
Solulan® 5	0.5
Mineral Oil (70 visc.)	2.0
Water Phase:	
BioCare™ Polymer HA-24 (2.6%)	3.8
Glucam® E-10	3.5
Water	80.2
Germaben® IIE	1.0

Add water phase (minus the **BioCare™ Polymer HA-24** and **Germaben® IIE**) at 85 C to the oil phase at 85 C with mixing. Add the **Germaben® IIE** at 75 C. Continue to mix to 40 C. Add the **BioCare™ Polymer HA-24** with mixing while cooling to 30 C.

Hand and Body Lotion

Formula No. 1

(Herbal)

A	**Cutina MD**	10.00
	Eumulgin® B-2	3.00
	Eutanol G	10.00
B	Propylene Glycol	5.00
	Water	68.85

C	**Hexaplant Richter**	3.00
	Fragrance	0.15
	Dyes	q.s.
	Preservatives	q.s.

Heat Part A to 70–75 C. Heat Part B to 70–75 C. Add Part B to A under agitation. Take heat off and continue stirring. At 40 C add individual components of Part C. Continue stirring and fill off at room temperature.

No. 2

Distilled Water	87.8
Sipon® LSB	3.0
Glydant®	0.2
Aldo® MSC Beads	2.0
Sipoteric C-20	1.0
Neo-Fat® 18-55	2.5
CO-1695	0.5
Mineral Oil	1.0
Glucamate® SSE-20	2.0
Fragrance, Color	q.s.

Add water, **Sipon®**, and **Glydant®** and begin heating to 65–70 C with moderate agitation. In a separate container, add **Aldo® MSC** beads, **Sipoteric C-20, Neo-Fat® 18-55, CO-1695**, mineral oil, and **Glucamate®** and heat to 65 C. When both phases have reached proper temperature, add oil phase to water phase, slowly, with moderate agitation. At 45–50 C add appropriate fragrance and color.

No. 3

A	Deionized Water	72.15
B	Polysorbate 20	2.00
	Methyl Paraben	0.20
	Ethyl Paraben	0.15
C	Glycerin	3.00
	Keltrol®	0.12

D	Safflower Oil	4.00
	Sesame Oil	4.00
	Mineral Oil	1.00
	Arlacel® 165	3.00
	Arlacel® 60	2.00
	Cetyl Alcohol	1.00
	Amerchol L-101®	3.00
	Dow Corning® 200 Fluid	0.50
	BHA	0.05
	Methocel® 40-100	0.10
	Vitamin E Oil	0.01
	Stearic Acid	1.50
E	Deionized Water	1.00
	Dowicil® 200	0.10
F	Deionized Water	1.00
	Collagen	0.01
	Elastin	0.01
G	Fragrance (floral)	0.10

Meter water (Part A) into a compounding vessel and heat to 75–80 C. Combine Part B ingredients and warm to dissolve. Add to water phase (Part A) at 75–80 C. Combine Part C ingredients and mix well to prewet xanthan. Allow the mixture to set for 15 min, then add to Part A batch at 75–80 C. Weigh Part D ingredients into a steam-jacketed kettle. Heat to 80 C while mixing. Then add to water phase (at 75–80 C) with good agitation. Turn heat off and mix emulsion down to 45 C. Prepare Part E by dissolving **Dowicil® 200** antimicrobial in water. Then add to batch at 45 C (or lower). Add Part F (dissolved collagen and elastin) at 45 C (or lower). Add fragrance (Part G) to the emulsion at 45 C (or lower).

Note: If color is desired in the product, an aqueous color solution can be added to the emulsion at about 60 C.

No. 4

(Orange Roughy)

Super Refined Orange Roughy Oil	10.00
Polawax®	5.00
Crodamol IPM	5.00
Propylene Glycol	5.00
Carbopol® 941	0.10
Triethanolamine	0.10
Methyl Paraben	0.11
Propyl Paraben	0.03
Germall® II	0.20
Distilled Water	74.46

Combine **Orange Roughy Oil, IPM,** and **Polawax®**, heat to 75 C. Mix water and propylene glycol and dust in the carbomer. When thoroughly dissolved add the preservatives while heating to 75 C. At 75 C add the TEA to the water phase. When both are heated add the water to the oil. Mix to room temperature. pH 6.6.

Dry Skin Lotion for the Shower

A	**Hetoxol CS**	3.5
	Parapel HC	2.5
	Arosurf TA-100	1.0
	White Petrolatum	2.0
	Carnation Mineral Oil	2.0
	Hetester PMA	2.0
B	Deionized Water	86.6
	Cellosize® QP (30,000 hi)	0.3
C	**Kathon CG**	0.1

Heat Part A to 85 C and mix well. Heat Part B to 85 C and disperse. Add Part A to Part B at 85 C and mix well. Continue to mix well with cooling to 45 C and add Part C. Avoid aeration. Continue to mix and cool to 35 C.

Moisturizing Lotion

Formula No. 1

A	**Cutina GMS**	5.00
	Eumulgin® B-1	1.00
	Eumulgin® B-2	1.00
	Cetiol 868	10.00
	Avocado Oil CLR	1.00
B	Water	76.70
	Sorbitol	5.00
	Dowicil® 200	0.10
C	Fragrance	0.20

Heat Part A to 75–80 C. Heat Part B to 75–80 C. Add Part B to A under agitation. Remove heat and continue mixing. At 45 C add Part C. Continue mixing until product reaches room temperature. Fill off.

No. 2

A	Mineral Oil	3.00
	Isopropyl Myristate	1.00
	Stearyl Alcohol	2.00
	Brij® 78	2.00
	Arlacel® 60	1.00
B	Water	50.00
	Carbopol® 940	0.05
C	**Arlatone PQ 225**	10.00
	Water	30.40
	Triethanolamine	0.05
D	**Germaben® II**	0.50

Disperse **Carbopol® 940** in water and heat Part B to 75 C. Heat Part A to 70 C. Add Part A to B and mix using propeller-type agitation. Heat Part C to 70 C and add to Parts A and B with thorough agitation. Remove heat, continue agitation and cool to 60 C. Add Part D. Cool to set point with moderate agitation.

Physical Properties:

Typical visc. — 7,000 cps
Stability — 3 months at 25 C, 40 C, and 50 C

No. 3

A	Petrolatum	2.0
	Arlamol E	2.0
	Stearyl Alcohol	1.0
	Brij® 72	3.0
	Brij® 78	1.0
	Silicone Fluid	0.1
B	Water	50.0
	Carbopol® 940	0.1
C	**Arlatone PQ 220**	10.0
	Water	30.2
	Triethanolamine	0.1
D	**Germaben® II**	0.5

Disperse **Carbopol® 940** in water and heat Part B to 75 C. Heat Part A to 70 C. Add Part A to B and mix using propeller-type agitation. Heat Part C to 70 C and add to Parts A and B with thorough agitation. Remove heat, continue agitation and cool to 60 C. Add Part D. Cool to set point with moderate agitation.

Physical Properties:

Typical visc. — 13,000 cps
Stability — 2 months at 25 C, 40 C, and 50 C

No. 4

Oil Phase:	
Amerlate® P	0.5
Stearic Acid	3.0
Glyceryl Stearate	2.0
Water Phase:	
Glucam® E-20	5.0
Triethanolamine	1.0
Water	83.7
BioCare™ Polymer HA-24 (2.6%)	3.8
Germaben® IIE	1.0

Add **BioCare™ Polymer HA-24** to water at 25 C. Mix until thoroughly dispersed. Heat to 45–50 C if necessary. When dispersed, add the **Glucam® E-20** and triethanolamine and heat to 85 C. Add the water phase at 85 C to oil phase at 85 C with good stirring. Add the preservative **Germaben® IIE** at 75 C. Continue to mix to room temperature with thorough agitation.

No. 5

A	**Lexemul® 561**	2.00
	Lexol IPM	1.00
	Lexemul® EGMS	0.50
	Mineral Oil	0.50
	Acetylated Lanolin Alcohol	1.00
	Cetyl Alcohol	1.00
	Stearic Acid	1.00
	Cyclomethicone	0.50
	Lexgard P	0.05
B	Sorbitol (70%)	4.00
	Magnesium Aluminum Silicate	0.30
	Triethanolamine	q.s. to pH 7
	Carbomer 934	0.20
	Lexgard M	0.15
	Water	q.s. to 100

Heat water to 70 C. Disperse all ingredients except Carbomer 934 and TEA. Mix thoroughly. Slowly disperse Carbomer 934, then neutralize. Melt contents of Part A together. Slowly add Part A to Part B with agitation. Readjust pH. Cool to set point and fill.

Physical Properties:

pH — 7.0
Visc. (@ 24 C) — 10,000 cps
Stability after one month — Stable at room temperature, at 45 C, and passes 3 freeze/thaw cycles

No. 6

(In-Shower)

A	**Cetiol LC**	1.50
	Myritol 318	4.00
	Mineral Oil	7.50
	Emulgade F	6.00
B	Water	77.70
	Cosmedia Polymer HSP-1180	3.00
C	Triethanolamine (for pH adjustment)	0.30
	Fragrance, Preservative, and Dyes	q.s.

Mix and heat Part A to 75 C. Mix and heat Part B to 75 C and add to Part A. Cool to 45 C and add individual components of Part C. Continue stirring until product reaches room temperature. Fill off.

Tonic Lotion

Formula No. 1

Vegetol® Lp	2.5
Vegetol® Cornflower (water soluble)	2.5
Kathon CG	0.1
Vegetol® Red Poppy GR 106 (water soluble)	1.0
Demineralized Water	93.9

Mix all the components at room temperature.

No. 2

(Body)

A	**Tefose® 1500**	8.0
	Geleol®	1.0
	Lanolin Wax	0.5
	Propylene Glycol Dipelargonate	5.0
B	Demineralized Water	40.0
C	Demineralized Water	36.1
	Kathon CG	0.1
	Vegetol® Horse Chestnut (water soluble)	4.0
	Vegetol® Witch Hazel (water soluble)	4.0
D	Essential Oils	1.0
	Perfume	0.3

Parts A and B are heated separately to 75 C. Add Part B slowly to Part A. Stir slowly for 5 min, then begin cooling. At about 30 C add Part C. Add Part D. Homogenize if necessary.

No. 3

(Greasy Skin)

A	Magnesium Chloride	2.0
	Rose Water Extract	42.4
	Hamamelis Water Extract	35.0
B	Human Placental Protein	10.0
	Kathon CG	0.1
	Alcohol	10.0
C	Perfume	0.1
	Solvating Agent γ 2420	0.4

At room temperature mix the components of Part A and mix the components of Part B. Add Part B to Part A, then add this blend to Part C.

Camphor Lotion

A	Camphor	0.50
	Propylene Glycol	40.00
	Vegetol® Yeast	5.00
B	Demineralized Water	53.35
	Perfume (water soluble)	0.05
	Dehydroacetic Acid (sodium salt)	0.10
	Solvating Agent γ 2420	1.00

At room temperature, mix the components of Part A, mix the components of Part B. Add Part B to Part A.

Astringent Lotion

A	Camphor	0.1
	Alcohol	25.0
B	Glycerin	0.6
	Vegetol® Rosemary	5.0
	Kathon CG	0.1
	Demineralized Water	68.9
	Perfume	0.3

At room temperature, mix the components of part A and mix the components of Part B. Add Part B to Part A.

Slimming Lotion

Vegetol® Seaweed (water soluble)	20.0
Vegetol® Birch-Tree (water soluble)	20.0
Vegetol® Horse Chestnut (water soluble)	20.0
Vegetol® Ivy (water soluble)	20.0
Vegetol® Hydrocotyle (water soluble)	19.9
Kathon CG	0.1

At room temperature, mix all the components. Filter the obtained blend.

Softening Lotion

A	Demineralized Water	84.1
	Vegetol® Mint (water soluble)	3.0
	Vegetol® Rosemary (water soluble)	2.0
	Fenugreek Glycolic Extract	5.0
B	Glycerin	5.0
	Kathon CG	0.1
C	Perfume	0.2
	Solvating Agent γ 2420	0.6

Mix the components of Part A then add glycerin and preservative. Add Parts A/B to Part C.

Beauty Lotion

A	**Lexemul® 561**	2.00
	Lexol IPM	4.00
	Stearic Acid	2.00
	Cetyl Palmitate	1.00
	Castor Oil	0.50
	Mineral Oil	0.50
	Lexgard P	0.05
B	Carbomer 934	0.20
	Sodium Hydroxide (50%)	q.s. to pH 7.5
	Lexgard M	0.15
	Water	q.s. to 100.00

Heat water to 70 C. Dissolve **Lexgard M**. Disperse Carbomer 934 and neutralize. Separately melt contents of Part A together. Slowly add Part A to Part B. Readjust pH, cool, and fill.

Physical Properties:

pH — 7.2
Visc. (@ 24 C) — 8,000 cps
Stability after one month — Stable at room temperature at 45 C and passes 3 freeze/thaw cycles.

Collagen Lotion

(Pourable)

A	**Veegum®**	1.00
	Rhodigel 23	0.50
	Water	76.25
B	Glycerin	6.00
	Collasol	5.00
	Allantoin	0.25
C	**Polysynlane**	4.00
	Amerchol L-101®	3.00
	Super Sterol Ester	2.00
	Crodafos N-3 Neutral	1.00
	Dow Corning® 200 Fluid (350 cs)	1.00
D	Preservative	q.s.

Blend **Veegum®** and **Rhodigel.** Slowly add to the water while agitating at maximum available shear. Continue mixing until smooth. Add Part B ingredients in order, allowing each addition to become smooth and uniform. Combine Part C ingredients in order, stirring until uniform. Add Part C to A/B. Mix until homogeneous. Add Part D.

Suggested Packaging: Glass or plastic bottle.

Note: The **Veegum®**/xanthan gum mixture stabilizes the emulsion and adjusts the viscosity of this cold process lotion. This lotion provides an elegant feel in both initial application and in the residual film. The final pH should be approximately 6.5.

Baby Lotion

Formula No. 1

A	**Drakeol 7**	26.00
	Penreco Snow	12.00
	White Beeswax	3.00
	Sorbitan Monostearate	3.00
	Polysorbate 60	4.00
	Propyl Paraben	0.15

B	Water	43.60
	Methyl Paraben	0.15
	Glycerin	8.00
	Suspending Agent	0.10

Disperse the suspending agent in the water at 70 C, then add the remaining ingredients in Part B. Mix the materials in Part A and heat them to 75 C. Blend Part A into Part B and allow to cool with continuous agitation. Perfume may be added at 45 C.

No. 2

A	**Drakeol 7**	37.50
	Penreco Snow	7.50
	White Beeswax	2.00
	PEG-40 Sorbitan Lanolate	4.50
	Sorbitan Sesquioleate	2.00
	Propyl Paraben	0.15
B	Water	46.20
	Methyl Paraben	0.15

Heat the ingredients in Part A to 70 C and those in Part B to 75 C. Add Part B to A while stirring and continue to agitate until the product cools to room temperature. Perfume may be added to the mixture after it has cooled to 45 C.

Note: Formula No. 1 produces a white opaque lotion with a silky feel. The lotion goes onto the skin smoothly and leaves a protective film. Formula No. 2 is basically similar but with a slightly heavier protective film.

No. 3

A	Demineralized Water	84.25
	Kathon CG	0.10
	Allantoin	0.10
	Glycerin	2.00
	Vegetol® Camomile (water soluble)	10.00
	Propylene Glycol	2.00
	Tetrasodium EDTA	0.05

B	Perfume	0.30
	Solvating Agent γ 2428	1.20

Add Part A to Part B at room temperature.

No. 4

A	**Tefose® 1500**	6.0
	Stearic Acid	1.0
	Sweet Almond Oil	5.0
	Mineral Oil	5.0
B	Demineralized Water	66.4
	Carbomer 934 (1% gel)	10.0
	Kathon CG	0.1
	Glycerin	3.0
C	**Vegetol® Marigold** (water soluble)	3.0
	Perfume	0.5

Parts A and B are heated to 75 C. Cool while stirring slowly. At about 25 C, add **Vegetol**. Add perfume as required. Homogenize if necessary.

No. 5

(Cationic)

Oil Phase:	
Carnation Mineral Oil	10.0
Polawax®	3.0
Fluilan	3.0
Crodacol C-95	2.0
Incroquat Behenyl tMC	1.0
Water Phase:	
Propyl Paraben	0.1
Glycerin USP	3.0
Methyl Paraben	0.2
Germall® 115	0.3
Deionized Water	77.4

Combine oil phase and heat to 80 C. Combine water phase and heat to 82 C. Slowly add water phase to oil phase under controlled agitation. Cool. Maintain mixing.

All-Purpose Cream

A	**Glucate® SS**	1.5
	Coconut Oil (76°)	7.5
	Glyceryl Monostearate (neut.)	7.5
B	**Glucamate® SSE-20**	1.5
	Glucam® E-20	5.0
	Water	76.0
	Germaben® II	1.0

Heat Part A and Part B separately to 75 C. Mix Part A until uniform. Add Part A to Part B with constant stirring and cool to room temperature.

Night Cream

Formula No. 1

A	**Lexemul® LPM**	5.00
	Lexemul® 561	5.00
	Lexol IPM	5.00
	Stearic Acid	2.00
	Mineral Oil	5.00
	Lexgard P	0.05
B	Glycerin	3.00
	Carbomer 934	0.15
	Triethanolamine	q.s. to pH 7.0
	Lexgard M	0.15
	Water	q.s. to 100.00

Heat water to 70 C. Add Carbomer 934 and neutralize, then add glycerin and **Lexgard M.** Separately melt contents of Part A together. Slowly add Part A to B with agitation. Mix until set point and fill.

Physical Properties:

pH — 7.0

Visc. (@ 24 C) — 35,000 cps

Stability after one month — Stable at room temperature at 45 C and passes 3 freeze/thaw cycles.

No. 2

(Foundation)

A	White Mineral Oil (USP)	6.000
	Cetyl Alcohol	10.00
	Avanel S-30	2.00
B	Propylene Glycol	5.00
	Methyl Paraben	0.25
	Deionized Water	76.75
	Color, Perfume	as desired

Heat Part A to 70 C, Part B to 75 C. Add Part A to B with moderate to high agitation, including side-scraping motion. Mix for a few minutes, then cool rapidly to 30 C. Emulsion will invert to give a viscous water-in-oil emulsion.

Note: The product is a nonpourable viscous cream with a pH about 6.8.

No. 3

(Aloe Vera)

A	Deionized Water	49.225
	Tetrasodium EDTA	0.075
	Propylene Gllycol	3.500
	Methyl Paraben	0.200
B	**Adol 52 NF**	2.000
	Ritachol® 1000	2.000
	Emersol® 132	4.000
	Polysorbate 40	2.000
	Sorbitan Palmitate	0.700

	Ritachol®	1.000
	Mineral Oil	7.000
	Aloe Vera Oil	3.000
	Ritaderm	3.000
	Dimethicone 200	1.000
	BHA	0.100
	Propyl Paraben	0.100
C	Sodium Borate	0.200
D	Aloe Vera Gel	20.500
	Fragrance	0.150
	Germall® II	0.25

Heat Part A and Part B separately with agitation to 75 C. Add Part A to Part B and mix 30 min. Add Part C and cool with agitation until temperature reaches 50 C. Add Part D and agitate until temperature reaches room temperature.

No. 4

A	**Silsoft™ Beauty Aid MG**	30.00
	Stearic Acid	3.50
	Stearyl Stearate	2.00
	Glycerol Monostearate SE	0.50
B	**Ucare® Polymer LR-400** (1% aq. sol'n.)	63.00
C	**Germall® II**	1.00
	Perfume	q.s.

Weigh and heat Parts A and B separately to 55 C (131 F). Add Part B slowly to Part A with good mixing under moderate shear. Continue mixing, and cool to 40 C (104 F). Add Part C, mix, and cool to 25 C (77 F).

No. 5

(Replenishing)

A	Water	81.4
	Germaben® II	0.2
	Monaquat P-TS	3.0
B	**CO-1695**	5.0
	Ceraphyl 424	5.0
	Wickenol® 101	4.0
	Ritawax YD503	0.4
	Dow Corning® 200 Fluid	1.0

Heat Part A to 65 C. Mix until uniform. Heat Part B to 65 C. Mix until uniform. Add Part A to Part B with stirring. Cool to 40 C and add fragrance, coloring, or preservative as required.

Physical Properties:

Appearance — Soft shiny cream, nonflowable
pH (as is) — 6.6

No. 6

(ATP Nucleotide)

Oil Phase:

Crodamol CP	5.0
Super Refined Orange Roughy Oil	10.0
Super Refined Shark Liver Oil	10.0
Polawax®	5.0
Syncrowax BB4	2.0
Fluilan	5.0

Water Phase:

Deionized Water	51.9
Incromectant AMEA-70	5.0
Carbopol® 941	0.2
Triethanolamine	0.2
Methyl Parben	0.2
Germall® 115	0.3

C	ATP Nucleotides	5.0
	Fragrance K5278	0.2

Heat oil phase to 78 ± 2 C. Dissolve **Carbopol®** in water phase. Heat water phase to 78 ± 2 C. Add water phase to oil phase. Cool. At 45 C add Part C.

Cold Cream

Formula No. 1

Water	49.6
Drakeol 9	33.0
White Beeswax	16.5
Borax	0.9

Mix all ingredients together and heat to 70 C. Agitate while cooling to room temperature. Add perfume at 45 C.

No. 2

A	**Ross Beeswax Substitute 628/5**	11.0
	Ross Fully Refined Paraffin 150/160	2.0
	Mineral Oil 80/90	47.5
	Glycerol Monostearate SE	0.3
B	Borax	0.8
	Water	38.4

Heat Part A to 170 F and blend, heat Part B to 170 F and blend. Cool to 160 F and add Part A to B at 160 F with good agitation. Cool slowly with agitation and pour at 110 F.

Placenta Skin Cream

A	**Lanette 16**	5.00
	Cetiol 1414E	5.00
	Eumulgin® B-1	0.40
	Eumulgin® B-2	0.35

B	Water	83.95
	Sorbitol	3.00
	Dowicil® 200	0.10
C	Fragrance	0.20
D	Placental Liquid (water soluble)	2.00

Heat Part A to 75–80 C. Heat Part B to 75–80 C. Add Part B to A under agitation. Cool to 45 C and add Part C. At 35 C add Part D. Continue mixing until product reaches room temperature. Fill off.

Vitamin E Cream

A	**Cutina GMS**	4.00
	Lanette O	1.50
	Eumulgin® B-1	0.50
	Eumulgin® B-2	0.50
	Cetiol S	8.00
	Covitol 1100	1.00
B	Water	79.20
	Glycerin	5.00
	Dowicil® 200	0.10
C	Fragrance	0.20

Heat Part A to 75–80 C. Heat Part B to 75–80 C. Add Part B to A under agitation. Remove heat and continue mixing. At 45 C add Part C. Continue mixing until product reaches room temperature. Fill off.

Collagen Cream

A	Glyceryl Monostearate	8.0
	Eumulgin® C-700	4.0
	Standamul® G	10.0
	Mineral Oil	20.0

B	Water	54.0
C	**Collagen CLR**	3.0
	Germaben® II	1.0

Heat Part A to 75 C and stir until uniform. Heat Part B to 75 C. Add Part A to B with constant stirring and stir to 35 C. Add Part C and continue stirring to room temperature.

Vanishing Cream

A	**Rosswax 63-0412**	8.0
	Rosswax 573	12.0
	Amerlite P	1.0
	Emerest® 2314	1.0
	Emerest® 2316	1.0
	Glyceryl Monostearate SE	0.5
B	Water	99.0
	Emery® 916 (pure)	8.0
	Triethanolamine	1.2
	Fragrance	q.s.

In separate steam-jacketed kettles heat both Part A and B to temperature to 170 F with agitation. When the temperature is reached add Part A to B with continued agitation, cooling to 120 F to package. Fragrance may be added to the product as it is cooling.

Hand Cream

A	**Tefose® 2561**	18.0
	Isostearyl Isostearate	6.0
	Phenyl Dimethicone	1.0
	Glyceryl Stearate	1.0
	Lanolin Wax	1.0
B	**Kathon CG**	0.1
	Demineralized Water	67.7

C	**Vegetol® Lemon GR 20** (water soluble)	5.0
	Perfume	0.2

Heat Parts A and B to 75 C. Add Part B to A. Cool while stirring slowly. At about 25 C, combine Parts A/B. Add perfume as required. Homogenize if necessary.

Hand and Body Cream

Formula No. 1

(All-Purpose)

A	White Mineral Oil (USP)	6.00
	Cetyl Alcohol	10.00
	Avanel S-150	4.00
B	Propylene Glycol	5.00
	Methyl Paraben	0.25
	Deionized Water	74.75
	Color, Perfume	as desired

Heat Part A to 70 C, Part B to 75 C. Add Part A to B with moderate to high agitation, including side-scraping motion. Mix for a few minutes, then cool rapidly to 30 C. Emulsion will invert to give a viscous water-in-oil emulsion.

Note: The product is a nonpourable viscous cream with a pH about 6.8.

No. 2

A	**Lexemul® 561**	2.00
	Lexate IL	0.50
	Mineral Oil	2.50
	Lexol PG 900	3.50
	Stearic Acid (triple-pressed)	4.50
	Cetyl Alcohol	1.00
	Lexgard P	0.05
B	Triethanolamine	0.80
	Propylene Glycol	3.50

Lexgard M	0.15
Deionized Water	q.s. to 100.00

Heat water to 70–75 C. Add remainder of ingredients in Part B and mix with agitation until dispersed. Melt ingredients of Part A together. Add Part A to B with agitation. Cool with mixing to 35–40 C.

Physical Properties:

pH — 7.0

Visc. (@ 24 C) — 12,500 cps

Stability after one month — Stable at room temperature @ 45 C and passes 3 freeze/thaw cycles.

Protective Face and Body Cream

A	Cocoa Butter	5.00
	Amerchol® CAB	3.00
	Stearic Acid (triple-pressed)	5.00
	Glycerol Stearate	8.50
	Amerscreen P	5.00
	Glucamate® SS20	2.00
	Propyl Paraben	0.10
B	**Veragel Liquid 1:1**	50.00
	Water	17.50
	Glucam® E-10	1.00
	Merquat® 100	1.50
	Triethanolamine	1.00
C	Methyl Paraben	0.10
	Germall® II	0.30
	Fragrance	q.s.
	Color (optional)	q.s.

Heat Part A and B separately to 75 C. Add Part B to A with mixing. Let cool to approximately 45–50 C. Add preservative system and fragrance.

Note: SPF is approximately 8–10.

Moisturizing Cream

Formula No. 1

(With Collagen)

A	**Cutina E-24**	4.00
	Cutina GMS	5.00
	Myritol 318	6.00
	Lanette O	2.00
	Cetiol V	6.00
B	Water	63.50
	Cosmedia Polymer HSP-1180	5.00
C	Triethanolamine (99%)	0.50
	Collagen CLR	3.00
	Fragrance, Preservative, and Dye	q.s.

Heat Part A to 70–75 C. Add Part B to A under agitation. Take heat off and continue stirring. At 40 C add individual components of Part C. Continue stirring until product reaches room temperature. Fill off.

No. 2

(Vegetable-Based)

Oil Phase:	
Super Refined Almond Oil	20.0
Crodesta F10	3.0
Syncrowax HGLC	2.0
Water Phase:	
Hydrosoy 2000 SF	1.0
Crodesta F110	3.0
Sorbo 70	5.0
Germaben® II	1.0
Veragel 200	0.5
Deionized Water	64.5

Combine oil phase. Heat to 78 ± 2 C. Combine water phase. Heat to 78 ± 2 C. Slowly add water phase to oil phase under agitation. Cool and fill off.

No. 3

A	**Carbopol® 934** (2% aq. sol'n.)	10.0
	Deionized Water	32.6
B	Liquid Paraffin (70)	25.0
	Jojoba Wax	10.0
	Ceresin	2.0
	Beeswax	8.0
	Lanolin Wax	1.5
	Glyceryl Stearate	3.5
	Ceteth-16	2.0
C	Triethanolamine (99%)	0.43
	Deionized Water	5.0
D	Fragrance	q.s.

Combine the **Carbopol®** resin solution and water using low-shear agitation. Heat to 70 C. Heat all ingredients in Part B in another vessel to 70 C. Add Part B to A using moderate-shear agitation. Add Part C to above mixture, with stirring, and cool to 40 C. Add Part D to above mixture, stirring well.

Stability Data (@ R.T.):

pH	7.1 @ 24 h	7.1 @ 1 mo
Brookfield Visc.	84,400 cps @ 24 h	69,200 cps @ 1 mo

Dry Skin Cream

Formula No. 1

A	**Modulan®**	3.7
	Amerchol L-101®	4.2
	Isopropyl Myristate	2.7
	Sodium Stearate	10.0
	Glyceryl Monostearate (pure)	1.8
	Ross Spermaceti Wax Sub. #573	5.5
	Ross Jojoba Oil	1.8

B	Water	59.7
	Emery® 916	9.2
	Triethanolamine	1.4

Melt Parts A and B in separate vessels to 170 F under agitation. When temperature is reached, mix Part A to B and cool. Package in containers at below 120 F.

No. 2

A	**Quatrisoft™ Polymer LM-200**	0.2
	Water	36.1
B	Carbomer 940	0.2
	Water	39.8
	Triethanolamine (99%)	0.2
C	Propylene Glycol	3.0
	Glucamate® SSE-20	3.5
D	**Amerchol L-101®**	8.0
	Modulan®	1.0
	Glyceryl Monostearate (neutral)	2.0
	Glucate® SS	1.5
	Mineral Oil	4.5
	Perfume and Preservative	q.s.

Disperse **Quatrisoft Polymer LM-200** in water at room temperature with good agitation. When thoroughly dispersed, heat to 75 C until uniform. Mix Part B separately until uniform at room temperature. Add Part A to B, minus triethanolamine, with mixing. When thoroughly mixed, add the triethanolamine with mixing until gel-like. Heat Parts A/B and Part C to 75 C. Add Part C to A/B. Heat Part D to 75 C and add to Parts A/B/C. Cool to room temperature with mixing.

Skin Butter

(Oil, Anhydrous)

Super Refined Orange Roughy Oil	43.35
Syncrowax HGL-C	4.00
Syncrowax HRS-C	2.00
Procetyl AWS	7.50
Crodamol PMP	3.00
Super Refined Shark Liver Oil	1.00
A-C® Polyethylene 400A	4.00
Propyl Paraben	0.15
Siloxane 03314	35.00

Combine oils, polyethylene, paraben, waxes, and heat to 90 C under good agitation. Maintain temperature until all solids are completely melted. Cooling procedure: (1) for stiff gel, fast cool with agitation. (2) for soft gel, slow cool with agitation. When batch temperature reaches 45 C, add **Siloxane**. Cool to room temperature.

Baby Cream

Formula No. 1

A	**Drakeol 7**	16.00
	Penreco Snow	10.00
	Lanolin	17.60
	White Beeswax	5.00
	Sorbitan Monostearate	3.00
	Polysorbate 60	4.00
	Propyl Paraben	0.15
B	Water	39.00
	Sorbitol Sol'n.	5.00
	Methyl Paraben	0.15
	Citric Acid	0.10

No. 2

A	**Penreco Mineral Jelly 15**	20.00
	Glyceryl Monostearate	10.00
	Liquid Lanolin	25.00
	White Beeswax	5.00
	Propyl Paraben	0.15
B	Water	39.70
	Methyl Paraben	0.15

For both Formula No. 1 and No. 2:

Blend Part A at 70 C and Part B at 75 C. Slowly add Part B to A, remove from heat and continue to stir until the mixture cools to room temprature. Perfume, if desired, may be added when the blend has reached 40 C.

No. 3

(W/O)

A	**Base WL 2586**	15.00
	Sweet Almond Oil	6.00
	Mineral Oil	47.00
	Petrolatum	5.00
	Methyl Paraben	0.10
B	Carbomer 934 (1% gel)	10.00
	Demineralized Water	15.00
	Zinc Oxide	1.00
	Sodium Borate	0.75
C	Perfume	0.15

Heat Part A and B to 75 C. Add Part B to A quickly. Maintain in heating for 5 min, then cool while stirring slowly. Add Part C.

No. 4

A	**Tefose® 1500**	10.0
	Sweet Almond Oil	6.0
	Mineral Oil	4.0
	Labrafil® M 2130 CS	3.0
B	Demineralized Water	68.5
	Glycerin	3.0
	Fondix® G BIS	1.0
C	**Vegetol® Marigold** (water soluble)	4.0
D	Perfume	0.5

Heat Parts A and B to 75 C. Add Part B to A. Cool while stirring slowly. At about 25 C, add Part C. Add Part D as required. Homogenize if necessary.

Skin Conditioning Mousse

Formula No. 1

Quatrisoft™ Polymer LM-200	1.0
Propylene Glycol	2.0
Ethanol SD-40	15.0
Deionized Water	82.0
Preservative and Perfume	q.s.

Concentrate: Mix ethanol with water at room temperature. Add **Quatrisoft Polymer LM-200** and other ingredients with water/ethanol mixture using rapid stirring at room temperature until polymer is completely dissolved.

Aerosol Fill: Fill Aluminum mousse can and charge with **A-46 Propellant** using a weight ratio of 100 parts of product to 20 parts of propellant.

No. 2

A	**Solulan® 16**	1.5
	Glucam® P-10	1.5
	Solulan® 98	2.0
B	Deionized Water	78.5
	Quatrisoft™ Polymer LM-200	0.5
	Preservatives	1.0
C	SD Alcohol 40	15.0

Concentrate: Add **Quatrisoft Polymer LM-200** to agitating water at 25 C. Heat to 70 C and mix until dissolved. Add preservatives to complete Part B. In a separate container add Part A ingredients and heat to 70 C with agitation. Add Part B to A and mix until uniform. Cool to 40 C and add Part C. Cool to 25 C when uniform.

Aerosol Fill: Fill aluminum mousse can and charge with **A-46 Propellant** using 90% product and 10% propellant.

Mousse-Type Moisturizer

Purified Water	67.8
Kelco K1A96	0.2
Akypo RLM-45-N	5.0
Ajidew N-50	3.0
Glydant®	0.2
Stearic Acid	2.5
Siponic® C-20	3.0
Mineral Oil	3.0
Fragrance	0.3
Propellant A-46	15.0

Disperse **Kelco K1A96** in water at 65 C using moderate agitation. While maintaining agitation and temperature add **Akypo, Ajidew N-50,** and **Glydant®**. Having previously placed stearic acid, **Siponic® C-20**, and mineral oil in a separate container and having heated them to 65 C with moderate agitation, add this oil mixture to the water phase. After thoroughly mixing products—about 20 min—begin cooling and add fragrance at 45–50 C. Place emulsion in suitable pressure container and charge with **Propellant A-46.**

Body Mousse

Super Refined Mink Oil	0.75
Lanexol AWS	2.50
Crodacol C-95	0.25
Volpo 3	1.50
Carnation Mineral Oil	3.00
Incromate SDL	4.00
Aminofoam C	0.75
Crosilk Liquid	1.00
Crotein CAA SF	1.00
Glycerin	3.00
Propylene Glycol	12.00
Crotein HKP S/F	1.00
Deionized Water	67.75
Fragrance Q9879	0.50
Germaben® II	1.00

Heat the oil phase to 75 C, heat the aqueous phase to 80 C. Add the water phase to the oils. Continue stirring. Pressurize and charge.

Suggested Propellant: 90% Concentrate, 10% **Propellant A-31.**

Massage Oil

Solulan® PB 5	3.0
Dow Corning® Silicone #344	16.0
Emerest® 2314	13.0
Drakeol #9	31.0
Coconut Oil	31.0
Escalol® 507	3.0
Ross Jojoba Oil	3.0
Perfume Nova Rome DE 51	q.s.

Load all ingredients into a vessel. Warm slightly until clear under agitation and package.

Transparent Gel

(Emollient)

A	**Eumulgin® B-3**	13.00
	Cetiol HE	25.00
	Cetiol V	5.00
B	Deionized Water	56.75
	Dyes	q.s.
C	Fragrance	0.10
	Kathon CG	0.10

Heat Part A to 70–75 C. Heat Part B to 70–75 C. Add Part B to A under agitation. Continue stirring and at 55–60 C add individual components of Part C. Once the product is homogeneous, fill off.

Facial Cleanser

Formula No. 1

(Mild)

Water	19.95
Akypo RLM-100	10.00
Sipon® ESY/CT	23.00
Glycerin	3.00
Glydant®	0.20
Triethanolamine	2.35
Glycol Distearate	1.50
Fragrance	q.s.
Natrosol® 250 HR (2.5%)	40.00

Add water to suitable container equipped with agitator and heating facilities. Add **Akypo, Sipon® ESY**, glycerin, and **Glydant®** to water using moderate agitation and begin heating to 65 C. While product is warming, add triethanolamine. When product is at 65 C, add glycol distearate. Turn off heat when a uniform dispersion is obtained. Begin cooling and add the previously prepared **Natrosol®** solution to the batch. At 45–50 C add a suitable fragrance.

Note: A **Natrosol®** solution is prepared by first sprinkling **Natrosol®** powder onto the surface of moderately-to-rapidly stirring water at ambient temperature. Agitation at a moderate rate is continued and warming is begun. When 50 C is reached, the solution will be clear and viscous, making it suitable for use.

Physical Properties:

Visc. @ 25 C — 5900 cps
Solids — 22.93%
pH — 7.5
Appearance — Opaque liquid

No. 2

(Liquid)

A	Deionized Water	73.040
	Chamomile	0.010
	Aloe Vera Gel	0.010
	Allantoin	0.001
	Triethanolamine	0.020
	Dowicil® 200	0.200
B	**Methocel® 40-100**	1.500
C	Glycerin	3.000
D	TEA Lauryl Sulfate	15.000
E	Polysorbate 20	2.000
	Vitamin A, D_2 Oil	0.010
	Vitamin E Oil	0.010
	Laneth-10 Acetate	2.000
F	Lauramide DEA	3.000
	Perfume Oil	0.200

Meter water into main vessel and add other Part A ingredients. Mix until dissolved. Add **Methocel®** polymer (Part B) and mix until completely dissolved. **Methocel®** will dissolve quickly because the water phase is alkaline. Add glycerin (Part C). Add TEA lauryl sulfate (Part D). In a separate vessel, combine Part E ingredients. Heat to 40 C and mix thoroughly. Then add to batch. In a separate vessel, combine Part F ingredients. Heat to 40 C and mix thoroughly. Then add to batch. The final pH of the solution can be adjusted, if desired, with citric acid.

Aloe Vera Scrub

Formula No. 1

(Cream)

A	Deionized Water	65.95
	Aloe Vera Gel	1.00
B	Propylene Glycol	2.00
	Methyl Paraben	0.20
	Ethyl Paraben	0.15
C	Glyceryl Stearate SE	5.00
	Mineral Oil	5.00
	Safflower Oil	1.00
	Sesame Oil	1.00
	Squalane	1.00
	Wickenol® 163	1.00
	Stearic Acid	2.50
	Cetyl Alcohol	0.50
	Methocel® 40-100	0.50
D	Deionized Water	1.00
	Triethanolamine	1.00
E	Color	q.s.
F	Perfume Oil	0.10
G	Deionized Water	1.00
	Dowicil® 200	0.10
H	**Polyethylene 9A**	10.00

Meter water (Part A) into a compounding vessel, add aloe vera, mix and begin heating to 80 C. In a separate vessel, prepare Part B by heating propylene glycol. Add parabens and, when dissolved, add to water phase (Part A). Water must be above 60 C. Prepare Part C by combining all ingredients in a separate vessel and heating to 80 C. **Methocel®** polymer can be added with no prewetting. At 80 C combine with batch and mix for 5 min. Add Part D to batch and begin to cool. Add color (Part E). Add perfume oil (Part F) when batch is below 45 C. Add Part G (dissolved **Dowicil® 200** antimicrobial) when batch is below 45 C. Sprinkle in polyethylene beads and mix well.

No. 2

(Lotion)

A	Deionized Water	29.90
	Aloe Vera Gel	1.00
B	**Veegum®**	1.00
C	Propylene Glycol	3.00
	Methyl Paraben	0.20
	Ethyl Paraben	0.10
D	Glycol Stearate	5.00
	Sesame Oil	0.50
	Safflower Oil	0.50
	Methocel® 40-100	0.50
	Propyl Paraben	0.10
E	Sodium Lauryl Ether Sulfate	20.00
F	Sodium Lauryl Sulfate	18.00
G	Cocamide DEA	0.50
	Perfume Oil (herbal)	0.10
H	Color	q.s.
I	Cocamidopropyl Betaine	5.00

J	Deionized Water	1.00
	Dowicil® 200	0.10
K	**Polyethylene 9A**	13.50

Prepare Part A by metering water into compounding vessel, adding aloe vera and mixing. Then add **Veegum®** (Part B) and mix until dissolved. Begin heating to 80 C. Prepare Part C in a separate vessel by heating propylene glycol to 80 C. Then add parabens and mix until dissolved. When dissolved, add Part C to water phase (Part A). Water temperature must be above 60 C. When batch reaches 60 C, add Parts E, F, and G, mixing well between additions. (Allow 5 min between each.) Add color (Part H) to batch. Add cocamidopropyl betaine (Part I) to batch. When batch is below 45 C, add Part J (dissolved **Dowicil® 200** antimicrobial). Slowly sprinkle in polyethylene beads and mix into batch.

Troubled Skin Cleanser

Distilled Water	63.7
Citric Acid	0.1
Sipoteric 1398	10.0
Alconate L-3	20.0
Glucamate® DOE-120	3.0
Glydant®	0.2
Sipomide 1500	3.0
Fragrance, Color	q.s.

Dissolve citric acid in water while heating to about 50 C. Add **Sipoteric 1398** and **Alconate** with moderate agitation to obtain a satisfactory solution while maintaining temperature. Add **Glucamate®** and begin cooling. Add **Glydant®** and **Sipomide.** Add suitable fragrance and color.

Note: Viscosity may be raised by increasing concentration of **Sipomide 1500** or **Glucamate®.**

Physical Properties:

Appearance — Clear liquid
pH — 7.3
Visc. (#3 spindle @ 6 rpm, 25 C) — 800 cps

Pearled Hand Cleaner

Formula No. 1

(13.5% Solids)

Water and Preservative	83.75
Neodol® 25-3A	6.00
Monamine 779	7.00
Monamid® 1007	1.00
Ethylene Glycol Monostearate	2.00
Sodium Chloride	0.25–0.50

Physical Properties:

Visc. (pH adjusted to 7.2 with 0.25% salt) — 750 cps
Visc. (pH adjusted to 7.2 with 0.50% salt) — 4500 cps

No. 2

(16.5% Solids)

Water and Preservative	83.5
Monamine 779	15.0
Ethylene Glycol Monostearate	1.5

Physical Properties:

Visc. (pH adjusted to 7.1) — 1300 cps

Clear Skin Cleanser

Distilled Water	74.2
Sipon® LT-6	15.0
Sipoteric 1398	7.5
Sipomide 1500	3.0
Glydant®	0.2
Citric Acid	0.1
Fragrance, Color	q.s.

Dissolve **Sipon®** in water with moderate agitation. Add remaining ingredients singly with continued agitation to provide satisfactory solubility before each addition.

Physical Properties:
pH — 7.45
Visc. — 9000 cps

Liquid Soap

Jordanol SL-300	30.00
Jortaine CAB-35	7.00
Jordamide JT-128	3.00
Glycol Stearate	1.50
Sodium Chloride	0.25
Citric Acid	0.12
Perfume	as desired
Preservative	as desired
Water	balance

With agitation, add the sodium chloride and citric acid to the water. When these are completely dissolved, add the **Jordanol SL-300, Jortaine CAB-35,** and **Jordamide JT-128**. Heat the bath to 65–70 C and add the glycol stearate. When the glycol stearate is completely dispersed, cool to 35 C and add the remaining ingredients.

Physical Properties:
pH — 7.0
Visc. — 7500 cps

Hand Soap

Sipon® LSB	55.0
Monamate LNT-40	5.0
Monamid® 1089	5.0
Cerasynt IP	0.5
Germaben® II	0.3
Sodium Chloride	0.3
Water	33.9

Dissolve sodium chloride in water. Add other ingredients and heat to 60 C to melt. Cool while stirring. Fill containers. Full pearlizing will occur overnight.

Note: Adjust pH to 6.8 with phosphoric acid.

Waterless Hand Cleaner

Oil Phase:	
Deodorized Kerosene	39.0
Pale Oleic Acid	7.5
Neodol® 25-3	2.0
Water Phase:	
Monoethanolamine	0.8
Triethanolamine	2.6
Propylene Glycol	2.5
Glycerin	1.0
Lanolin	0.5
Butyl Oxitol®	0.4
Water	43.7
Perfume, Color	as desired

Heat both oil and water phases separately to 70–75 C. Pour oil into water while stirring contiuously. Continue until a homogeneous smooth gel is formed. Pour into jars. Let cool to room temperature.

Physical Properties:

Stable at 49 C (120 F) for 7 days and after five freeze-thaw.

Jojoba Soap Bar

Duveen Toilet Soap Base	93.55
Ross Powdered Jojoba Meal	5.00
Ross Jojoba Oil	0.50
Novarome Fragrance CD-69	0.75
Titanium Dioxide	0.20

Shampoo

Formula No. 1

Water	64.05
Sipon® LS	15.00
Witconate AOS Slurry	10.00
Hamposyl L-30	5.00

Hetsorb L-20	4.00
Germaben® II	1.00
Jaguar® HP-8	0.75
Alpine #137-045	0.20
Citric Acid	q.s.
Color	q.s.

To water add **Jaguar® HP-8.** To hydrate gum, adjust to pH 5–7 with citric acid. Add **Witconate AOS, Sipon® LS, Hamposyl L-30, Hetsorb L-20, Germaben® II,** and **Alpine #137-045**. Adjust pH to 5–6 with citric acid.

No. 2

(Extra Mild)

Water	55.00
Monamate CPA-40	20.00
PEG 6000 Distearate	1.00
Clindrol 100 CG	3.00
Chemadene NA-30	15.00
Peptein 2000	3.50
Preservative, Color, Perfume	q.s.
Citric Acid (to pH 6.3–6.8)	q.s.
Water	q.s. to 100.00

Heat the water to 60–65 C. Add the next three items. When the liquid is clear, cool to 140 C and add the **Chemadene.** Add the **Peptein 2000** at 35–40 C. Adjust pH.

No. 3

(Clear, Mild)

Water and Preservative	61.0
Monamate LNT-40	15.0
Monateric LMAB	20.0
Monamid® 1089	4.0

Add ingredients in order listed and blend thoroughly. No heat is

necessary. Adjust pH with 50% citric acid to desired level.
Note: At pH 6.5 viscosity is approximately 3000 cps.

No. 4

(All-Purpose)

Demineralized Water	63.75
Odorisol SPS	28.00
Amphosol CA	3.00
Ninol 128 Extra	1.00
Formaldehyde (30% aq. sol'n.)	0.20
Citric Acid (50% aq. sol'n.)	0.20
Tetrasodium EDTA	0.10
Vegetol® Camomille (water soluble)	1.00
Perfume	0.25
Sodium Chloride	2.50

All the components are mixed at room temperature.

No. 5

(Family)

Water	14.1
Monamate LNT-40	25.0
Monamid® 1089	5.0
Sipon® LSB	55.0
Cerasynt IP	0.5
Phosphoric Acid (85%)	0.4

Mix **Sipon® LSB**, water, and preservative. Add phosphoric acid (85%). Stir and add **Monamid® 1089**. Stir until homogeneous and add **Monamate LNT-40**. Add **Cerasynt IP**. Stir and heat to completely melt the solid material (approx. 60 C). Cool with stirring. At ≤ 40 C add any perfume or colors. Continue stirring and cooling to 25 C. Adjust pH to 6.0–7.0.

Physical Properties:

Physical Appearance — Pearly liquid
Nominal Activity — 31%
Visc. — 4000–5000 cps

No. 6

(Egg Lotion)

Sipon® LT6	42.0
Ethylene Glycol Monostearate (pure)	1.5
Lanolin	0.5
Spray-Dried Whole Egg	1.0
Deionized Water	55.0
Perfume	q.s.
Dye	q.s.

Combine **Sipon® LT6** and egg in water with agitation. Add remaining ingredients, except perfume and dye, and heat to 140 F with stirring. Allow mixture to cool to 100 F prior to addition of perfume and color.

No. 7

(Creamy Aloe)

Veragel Liquid 1:1	50.0
Deionized Water	23.0
Merquat® 550	3.0
Sodium Lauryl Sulfate	7.0
Ammonium Laureth Sulfate	5.0
Sodium Laureth Sulfate	5.0
Lauramide DEA	5.0
Sodium Chloride	1.0
Germaben® II	1.0

Add all ingredients and heat to 70 C with stirring.

No. 8

(Pearly)

A	**Odorisol SPS**	30.00
B	**Ninol 128 Extra**	3.00
	Glycol Stearate	3.00
C	Demineralized Water	53.45
	Vegetol® Rosemary (water soluble)	5.00
	Formaldehyde (30% aq. sol'n.)	0.20
	Citric Acid (50% aq. sol'n.)	0.20
	Perfume	0.15
	Amphosol CA	5.00

Parts A and B are heated to 60 C. Part B is added into part A while stirring, then Part C components are added.

No. 9

(Pearlized)

Ceraphyl 60	2.0
Cerasynt IP	3.0
Foamole A	3.0
Ammonyx LO	2.0
Water	36.5
Germaben® II	1.0
Maprofix® TLS-500	42.5
Sodium Chloride (20% aq. sol'n.)	10.0
Citric Acid (30% aq. sol'n.) (to pH 6.5 ± 0.5)	q.s.

Add all ingredients and heat to 80–85 C with agitation. Mix at that temperature until a clear uniform mixture is obtained. Start cooling with agitation to 25–30 C, adjusting pH at 40 C to 6.5 ± 0.5.

No. 10

(Pearlized, For Permanent Waved Hair)

Polawax®	3.00
Croquat M	0.50
Incronol ALS	15.00
Incromide CME	3.00
Incronam I-30	1.00
Palmitic Acid	2.00
Methyl Paraben	0.20
Propyl Paraben	0.10
Distilled Water	75.20
Citric Acid	to pH 4.8

Charge a vessel with the water and add in the parabens. Start heating to 70 C. When dissolved add the **Incronol ALS** and **Incronam I-30** and stir until clear. Add in the rest of the ingredients keeping heat at 70 C until well mixed. Cool with mixing to 40 C and add the **Croquat M.** Adjust pH to 4.8. (Shampoo will thicken after 24 h.)

Physical Properties:

Visc. @ 24 h
(Brookfield RVT Spindle #3, 20 rpm, 25 C) — 1700 cps ± 10%

No. 11

(Pearlescent)

Sodium Lauryl Sulfate	30.00
Lexaine C	6.00
Lexemul® EGMS	1.00
Polypeptide LSN	1.00
Sodium Chloride	1.50 or q.s. to desired visc.
Lexgard M	0.15
Lexgard P	0.05
Citric Acid	0.05
Water	q.s. to 100.00

Heat water to 70 C. Add ingredients slowly with agitation to insure complete dispersion. Adjust pH, cool to 40 C and fill. Full pearlescence develops within 24 h.

Physical Properties:

pH — 6.0
Visc. (@ 24 C) — 1625 cps
Foam Height (initial) — 260 ml
Foam Height (5 min) — 225 ml
Lather Characteristics — Small and large size bubbles
Stability After 1 Month — Stable at room temperature, at 45 C, and passes 3 freeze/thaw cycles

No. 12

(Low Cost)

A	Deionized Water	74.097
	Triethanolamine	0.050
B	**Methocel® 40-101**	1.400
C	Deionized Water	1.000
	Citric Acid	0.200
	Dowicil® 200	0.150
	DL Panthenol	0.001
	Hydrolyzed Animal Protein	0.001
	Aloe Vera Gel	0.001
D	Color	q.s.
E	Sodium Laureth Sulfate	20.000
F	Cocamide DEA	3.000
	Perfume Oil	0.100

Weigh Part A ingredients into a mixing vessel and mix for 10 min. Having made the water alkaline, **Methocel®** polymers (Part B) can be slowly sprinkled in and will dissolve with mixing after about 15 min. Combine and dissolve Part C ingredients. When dissolved, add to Part A batch and mix for

10 min. Add color (Part D) and mix for an additional 10 min. Slowly add sodium laureth sulfate (Part E) to the batch and mix for 15 min (or until the batch is homogeneous). Combine Part F ingredients and heat to 40 C. Then mix to solubilize the perfume oil. Add Part F solution to batch and mix for 10 min.

No. 13

(Low Cost)

Sodium C14-16 Olefin Sulfonate (40%)	28.00
Lexaine CS	6.30
Cocamide DEA	2.00
Ammonium Chloride	2.50
Lexgard M	0.15
Lexgard P	0.05
Phosphoric Acid	q.s. to pH 6
Water	q.s. to 100.00

Add ingredients to water at 70 C. Mix until clear, then fill.

Physical Properties:

pH — 6.0
Visc. (@ 24 C) — 3000 cps
Foam Height (initial) — 325 ml
Foam Height (5 min) — 310 ml
Lather Characteristics — Small size bubbles. Thick
Stability After 1 Month — Stable at room temperature, at 45 C and passes 3 freeze/thaw cycles

No. 14

(Professional-Type)

Distilled Water	38.7
Sodium Chloride	1.0
Citric Acid	0.1
Sipon® ESY	24.0
Sipon® LD	15.0

Lexein X-350	2.0
Sipoteric COB	9.0
Sipomide 843	5.0
Formalin	0.1
Dowicil® 200	0.1
Propylene Glycol	5.0
Fragrance, Color	q.s.

Dissolve sodium chloride and citric acid in water using moderate agitation. Add remaining ingredients, singly, continuing agitation to obtain a satisfactory uniform dispersion before making each addition.

Physical Properties:

Appearance — Clear liquid
pH — 7.85
Visc. (#3 spindle, 6 rpm) — 4600 cps

No. 15

(High-Peformance Gel)

Water and Preservative	29.3
Alpha Olefin Sulfonate (40%)	15.3
Monateric LMM-30	20.4
Monamate OPA-30	35.0

Add ingredients in order listed while blending. Adjust pH to level desired. Viscosity will increase as pH is lowered. Viscosity will also vary widely with various brands of AOS, and can be increased if needed by addition of ammonium chloride. At pH 5.5–7.0 viscosities of 20–35,000 can be obtained.

No. 16

(Clear Protein)

Sodium Lauryl Sulfate	30.00
Lexaine C	6.00
Lexein QX3000	3.00

Lexgard M	0.15
Lexgard P	0.05
Sodium Chloride	0.50
Phosphoric Acid	q.s. to pH 6
Water	q.s. to 100.00

Heat water to 70 C. Add first two ingredients slowly with agitation. Adjust pH, then add remaining ingredients at 55 C.

Physical Properties:

pH — 6.1
Visc. (@ 24 C) — 563 cps
Foam Height (initial) — 325 ml
Foam Height (5 min) — 300 ml
Lather Characteristics — Medium size bubbles. Thick.
Stability After 1 Month — Stable at room temperature, at 45 C and passes 3 freeze/thaw cycles

No. 17

(Protein)

Sodium C14-16 Olefin Sulfonate (40%)	30.00
Diethanolamine Lauryl Sulfate	9.00
Lauramide DEA	3.50
Lexaine C	3.50
Lexein X350	2.00
Sodium Chloride	1.40
Lexgard M	0.15
Lexgard P	0.05
Citric Acid	0.15
Water	q.s. to 100.00

Heat water to 65 C. Disperse ingredients and agitate until clear. Adjust pH, then fill.

Physical Properties:

pH — 5.0
Visc. (@ 24 C) — 11,100 cps
Foam Height (initial) — 325 ml

Foam Height (5 min) — 325 ml
Lather Characteristics — Small size bubbles. Very thick
Stability After 1 Month — Stable at room temperature, at 45 C and passes 3 freeze/thaw cycles

No. 18

(Balsam and Protein)

Distilled Water	41.6
Natrosol® 250 HR	1.0
Sipon® L-22	50.0
Lexein X-350	2.0
Hamp-Ol 120	0.1
Citric Acid	0.1
Glydant®	0.2
Sipomide 843	5.0
Fragrance, Color	q.s.

Heat water to 65 C and add **Natrosol®** with rapid agitation. When a uniform dispersion is obtained, discontinue heating. Add remaining ingredients, singly, with moderate agitation to obtain a uniform dispersion before making next addition. Add fragrance and color at 50 C or below.

Physical Properties:
Appearance — Clear liquid
pH — 8.1
Visc. (#3 spindle, 6 rpm) — 17,600 cps

No. 19

(Oily Hair)

Demineralized Water	47.90
Kathon CG	0.10
Odorisol SPS	20.00
Amphosol AA	10.00
Phylderm® Filatov Aqueux	10.00
Polysorbate 20	2.00

Vegetol® Common Nettle (water soluble)	5.00
Perfume	0.25
Solvating Agent γ 2428	0.50
Citric Acid (50% aq. sol'n.)	0.25
Ninol 128 Extra	4.00

Add all components to the water.

No. 20

(Conditioning)

Avanel S-30	40.00
Jortaine CAB-30	10.00
Jordamox LDA	10.00
Lauramide DEA	4.00
Methyl Paraben	0.25
Citric Acid	q.s. to pH 7.0
Perfume and Color	as desired
Deionized Water	q.s. to 100.00

Heat Lauramide DEA to 40 C. Add to rest of ingredients at 40 C. Cool to 35 C, add perfume and color. Cool to room temperature.

No. 21

(Pearly Conditioning)

Water	37.35
Sipon® L-22	50.00
Hexylene Glycol	0.10
Citric Acid	0.15
Ajidew N-50	1.00
Sipoteric 1442	5.00
Sipomide 1500	3.50
Sipoteric COB	2.80
Dowicil® 200	0.10
Fragrance, Color	q.s.

After adding water, add remaining ingredients singly with sufficient agitation to provide a uniform dispersion before making next addition.

Notes: In the laboratory, this product could be made at room temperature. Warming to 40–45 C would quicken solubility of some ingredients.

Viscosity may be increased by removing hexylene glycol or increasing **Sipomide 1500** or **Sipoteric COB.**

Physical Properties:

Appearance — Pearly white liquid
Visc. (@ 25 C) — 4300 cps
pH — 6.2
Solids — 20.80%

No. 22

(Pearlized Conditioning)

Miranol® 2MCA-ESF	45.0
Mirapol® AD-1	2.0
Super-Amide GR	2.0
Tween® 20	1.0
Cerasynt IP	1.0
Water	49.0

Mix and heat all ingredients. Stir until uniform. Adjust pH to 7.0 with citric acid.

Physical Properties:

Solids — 20.7%
Visc. — 1800 cps

No. 23

(Pearlescent Conditioning)

Mirataine® CBS	10.0
Miranol® C2M Conc. N.P.	10.0
Cedepal® TD 404M	7.0
Maprofix® ES	10.0

Ucare® Polymer JR 400 (2% sol'n.)	50.0
Cerasynt IP	0.5
Water	12.5

Prepare a 2% stock solution of **Ucare® Polymer JR 400** by dispersing 2.0 parts of **Ucare® Polymer JR 400** in 98 parts of mildly agitated water at 20–25 C. When the polymer is completely wetted, heat to 60–70 C and agitate for approximately one hour until the solubilization is complete. Allow to cool with stirring.

Mix together while stirring the **Ucare® Polymer JR 400** solution, **Mirataine® CBS**, and **Miranol® C2M Conc. N.P.** Add the water, **Maprofix® ES, Cedepal® TD 404M**, and **Cerasynt IP**. Heat and mix until uniform making sure that the **Cerasynt IP** has dissolved. Cool with stirring. Adjust the pH to 6.7 with hydrochloric acid.

Physical Properties:

Solids — 15.7%

Visc. — 4000 cps

No. 24

(Conditioning)

A	**Quatrisoft™ Polymer LM-200**	0.5
	Water	54.0
B	**Glucam® E-20**	3.0
	Solulan® 16	2.5
C	Sodium Laureth-2 Sulfate (25% active)	35.0
	Lauramide DEA	5.0
	Citric Acid	q.s. to pH 7.0
	Perfume and Preservative	q.s.

Add **Quatrisoft™ Polymer LM-200** to water with good mixing at room temperature. When thoroughly dispersed, begin heating to 45 C until completely hydrated. Add **Glucam® E-20**. When uniform add **Solulan® 16** and heat with moderate mixing to 45 C. Avoid air entrapment. When clear and uniform add Part C. Mix until uniform. Adjust to pH 7.0 with citric acid.

No. 25

(Conditioning)

Sodium Laureth Sulfate (60%)	32.00
Lexaine C	15.00
Lexein X250	2.00
Lexamine 0-13	3.00
Lexgard M	0.15
Lexgard P	0.05
Lactic Acid	q.s. to pH 4.0
Water	q.s. to 100.00

Disperse first two ingredients into water at 70 C. Adjust pH to 4.0. Slowly add **Lexamine 0-13**, maintaining the pH at 4.0. Add remaining ingredients, cool and fill.

Note: The viscosity can be increased by adding sodium chloride to the formula.

Physical Properties:

pH — 4.0
Visc. (@ 24 C) — 200 cps
Foam Height (initial) — 400 ml
Foam Height (5 min) — 380 ml
Foam Characteristics — Medium and small size bubbles. Thick.
Stability After 1 Month — Stable at room temperature, at 45 C and passes 3 freeze/thaw cycles

No. 26

(Keratin Therapy)

Deionized Water	39.80
Incronol TLS	30.00
Sodium Laureth Sulfate (2 mole E.O.)	10.00
Incromide LR	10.00
Hexylene Glycol	1.00
Tween® 20	0.20
Glyceryl Monostearate SE	3.00
Crotein K	1.00

Incronam 30	4.00
Germaben® II	1.00

Combine ingredients and warm gently under agitation until uniformity. Cool.

No. 27

(Dry Hair)

Amphoteric 2	2.5
TEA Lauryl Sulfate	3.2
Sodium Lauroyl Sarcosinate	4.5
Lauramide DEA	4.0
Sodium Chloride	1.0
Water, Perfume, Color, Preservative	q.s. to 100.0

Note: Adjust to pH 6.2 with citric acid.

No. 28

(Clear for Dry Hair)

Water	40.75
Sipon® L-22	53.50
Ammonium Chloride	0.50
Citric Acid	0.15
Prologen I	1.00
Sipomide 1500	4.00
Dowicil® 200	0.10
Fragrance, Color	q.s.

After adding water, add remaining components singly with sufficient agitation to obtain a satisfactory solution before making next addition.

Notes: In the laboratory, this product could be made at room temperature. Warming to 40–45 C would quicken solubility of some ingredients.

Viscosity may be increased by increasing ammonium chloride or **Sipomide 1500** content.

Physical Properties:

Appearance — Clear off-white liquid
Visc. (@ 25 C) — 3200 cps
pH — 6.3
Solids — 21.15%

No. 29

(For Dry/Damaged Hair)

Incromide LR	5.00
Incronam 30	12.00
Super Refined Soybean Oil	0.30
Super Refined Sesame Oil	0.30
Super Refined Orange Roughy Oil	0.30
Croquat M	1.00
Cosmowax J	4.00
Volpo 3	1.00
Crodafos SG	2.00
Standapol® ES-2	32.00
Germaben® II	1.00
Distilled Water	41.10

Combine water and **Germaben®** and mix until blended. Add the SLES, then the **Super Refined** oils. Heat to 75 C while adding the **Incromide** and **Incronam**; when blended add **Cosmowax, Volpo,** and **Crodafos** mixing to 80 C. Mix with cooling to 40 C and add the **Croquat**. Continue mixing to room temperature.

Physical Properties:

pH — 5.95–6.10

No. 30

(Opaque for Oily Hair)

Water	47.8
Sipon® L-22	28.5
Siponate AOS-A246-LX	17.5
Ammonium Chloride	1.0

Citric Acid	0.2
Opacifier E-305	0.4
Sipomide 1500	4.5
Dowicil® 200	0.1
Fragrance, Color	q.s.

After adding water, add remaining ingredients singly, with sufficient agitation to provide a uniform dispersion before making next addition.

Notes: In the laboratory, this product could be made at room temperature. Warming to 40–45 C would quicken solubility of some ingredients.

Viscosity may be increased by increasing the ammonium chloride or **Sipomide 1500** content.

Physical Properties:

Appearance — White opaque liquid
Visc. (@ 25 C) — 3600 cps
pH — 6.0
Solids — 20.86%

No. 31

(Chlorine Neutralizing)

Water	36.3
Polymer JR-30M	1.0
Citric Acid	0.1
Dowicil® 200	0.1
Propylene Glycol	5.0
Ethylene Glycol Distearate	1.0
Sipon® L-22	42.9
Sipoteric COB	8.6
Sipoteric 1442	5.0
Fragrance, Color	q.s.

Disperse **Polymer JR** in water by heating to about 50 C with moderate agitation. Add citric acid, **Dowicil® 200**, and propylene glycol with continued agitation to obtain a uniform mixture. Heat to 65 C and add ethylene glycol distearate with continued agitation to obtain a uniform mixture. Discontinue heating. Add **Sipon® L-22, Sipoteric COB,** and **Sipoteric 1442** singly with sufficient agitation to obtain a uniform mixture before making next addition. At 45–50 C add a suitable fragrance and color.

Physical Properties:
Appearance — White opaque liquid
Visc. (@ 25 C) — 10,700 cps
pH — 5.5
Solids — 23.65%

Baby Shampoo

Formula No. 1

Water	51.85
Sipex® EST-30	30.00
Sipoteric 1398	5.00
Sipoteric COB	5.70
Akypo RLM-160N	5.50
PEG-150 Distearate	1.50
Citric Acid	0.25
Dowicil® 200	0.10
Benzyl Alcohol	0.10
Fragrance, Color	q.s.

Physical Properties:
Appearance — Clear liquid
Visc. (@ 25 C) — 840 cps
pH — 6.0
Solids — 20.15%

No. 2

Sodium Laureth Sulfate (29%)	25.00
Lexaine CSB-50	6.00
Polysorbate 20	1.00
PEG-150 Distearate	2.0 or q.s. to desired visc.
Lexgard M	0.15
Lexgard P	0.05
Boric Acid	q.s. to pH 6.5–7.0
Water	q.s. to 100.00

Blend ingredients together at 70 C. Cool and fill.

Physical Properties:

pH — 7.0
Visc. (@ 24 C) — 725 cps
Foam Height (initial) — 250 ml
Foam Height (5 min) — 250 ml
Lather Characteristics — Small size bubbles. Very thick.
Stability After 1 Month — Stable at room temperature, at 45 C, and passes 3 freeze/thaw cycles

No. 3

(Opaque)

Water	64.3
Sipoteric 1398	15.0
Sipoteric 1304	15.0
Benzyl Alcohol	0.1
Citric Acid (20%)	1.1
Glycol Distearate	1.5
Glucamate® DOE-120	3.0

Physical Properties:

Visc. (@ 22 C) — 3800 cps
pH — 6.45

No. 4

(Protein)

Sodium Laureth Sulfate (25%)	24.00
Lexein S620	20.00
Lexaine CSB-50	5.80
PEG-150 Distearate	0.50 or q.s. to desired visc.
Lexgard M	0.15
Lexgard P	0.05
Citric Acid	q.s. to pH 6.5
Water	q.s. to 100.00

Heat water to 70 C. Add ingredients and blend until clear. Adjust final pH and fill.

Physical Properties:

pH — 6.5
Visc. (@ 24 C) — 31 cps
Foam Height (initial) — 275 ml
Foam Height (5 min) — 275 ml
Foam Characteristics — Medium size bubbles
Stability After 1 Month — Stable at room temperature, at 45 C, and passes 3 freeze/thaw cycles

Dandruff Shampoo

Formula No. 1

Water	39.9
Sodium Chloride	0.5
Sipon® ESY	40.0
Alconate CPA	12.5
Dowicil® 200	0.1
Zinc Omadine (48%)	2.0
Sipomide 843	5.0
Fragrance, Color	q.s.

Dissolve the sodium chloride in the water with moderate agitation. Add the remaining ingredients singly using sufficient agitation to provide a uniform mixture before making next addition. The **Sipomide 843** is added to the stirring product while in a molten state. Add a suitable fragrance and color.

Notes: In the laboratory this product was made at room temperature with the exception of the heat needed to melt the **Sipomide 843.**

The shampoo viscosity made be raised by increasing the sodium chloride concentration.

Physical Properties:

Appearance — White opaque liquid
Visc. (@ 25 C) — 3900 cps
pH — 7.6
Solids — 21.76%

No. 2

(Moisturizing)

Sodium Laureth Sulfate (30%)	25.00
Lexein S620	10.00
Cocamide DEA	1.00
Propylene Glycol	3.00
Zinc Pyrithone (48%)	4.50
Magnesium Aluminum Silicate	1.00
Hydroxypropyl Methyl Cellulose	1.20
Lexgard M	0.15
Lexgard P	0.0
Water	q.s. to 100.00

Heat water to 70 C. Disperse methyl cellulose and magnesium aluminum silicate. Add remaining ingredients and mix thoroughly. Cool to 40 C and fill.

Physical Properties:

pH — 6.7
Visc. (@ 24 C) — 2925 cps
Foam Height (initial) — 320 ml
Foam Height (5 min) — 310 ml
Lather Characteristics — Small size bubbles. Thick.
Stability After 1 Month — Stable at room temperature, at 45 C, and passes 3 freeze/thaw cycles

No. 3

(Clear)

Croquat M	1.00
Crodesta SL-40	5.00
Incromide LR	3.00
Incronol ALS	25.00
Omadine® MDS	1.50
Germaben® II	1.00
Deionized Water	63.20
Uvinol MS-40	0.30
Citric Acid	q.s. to pH 5.0
FD&C Blue #1	q.s.

Premix 10 g of water with the **Omadine® MDS**. Charge a vessel with the **Incronol ALS** and **Incromide LR**. Start to heat to 70 C while adding the rest of the water, **Crodesta SL-40**, and **Germaben.®** When well blended, cool with mixing to 40 C. Adjust pH to 5.0 with citric acid. Add the water-MDS premix at 35 C and mix until clear.

Note: It is the responsibility of the user to seek and obtain FDA approval for applications of **Omadine® MDS** prior to marketing.

No. 4

(Luxury, Conditioning)

Croquat M	1.00
Crodafos SG	1.00
Incronam 30	20.00
Incromectant AMEA-70	1.00
Standapol® ES-2	25.00
Omadine® MDS	1.50
Uvinol MS-40	0.30
Germaben® II	1.00
Deionized Water	49.20
FD&C Blue #1	q.s.

Premix 20% of water with the **Omadine® MDS.** Charge a vessel with the remaining water, **Germaben®, AMEA, Crodafos SG, Croquat M,** and **Uvinol MS-40.** When well mixed add the SLES. Stir until clear with heating to 70 C. Cool. At 35 C add the water-**Omadine®** premix slowly to prevent aeration. Stir carefully until clear.

Note: It is the responsibility of the user to seek and obtain FDA approval for applications of **Omadine® MDS** prior to marketing.

Physical Properties:
pH — 6.0

Medicated Shampoo

Water	57.2
Veegum® K	1.5
Tetrasodium EDTA	0.1

Monamid® S	1.5
Monamate CPA-40	12.5
Monateric CAB	12.5
Alpha Olefin Sulfoante (40%)	12.5
Preservative	0.2
Sulfur	1.0
Salicylic Acid	1.0

Add the **Veegum®** to the water slowly, agitating continually while warming gradually. Add **Monamid® S** and continue agitation and heating to 80 C, until the **Monamid® S** melts and the mixture thickens. Remove heat and add remaining ingredients in order listed. If more viscosity is needed, the level of **Monamid® S** can be increased to 2.0% or **Veegum® HV** substituted for **K.**

Physical Properties:

Appearance — Opaque, ivory white
pH (as is) — 4.7
Visc. — 4000 cps

Shampoo Concentrate

(1:7 Dilutable)

Sipon® LT 6	37.6
Sipon® LSB	5.3
Monamine ACO-100	21.5
Monamide 150-ADD	18.8
Monamide 15-70W	6.5
Sodium Xylene Sulfonate (40%)	2.2
Monaquest ICA-120	4.3
Hexylene Glycol (UCC)	3.8

Blend ingredients in the order listed at 50 C with agitation until homogeneous. Add perfume and dye as desired.

Powdered Shampoo

Witconate AOK	35.20
Lanthanol LAL	35.10
Jaguar® HP-8	23.00
Jaguar® C-13-S	5.00
Crotein Q	1.00
Citric Acid	0.50
Methyl Paraben	0.10
Zero Free	0.10
Fragrance	q.s.
Color	q.s.

Dry blend ingredients.
Use level: 3.75%

Physical Properties:
pH — 5.5
Visc. (RV-4, 20 rpm) — 1950 cps

Aerosol Shampoo

	Formula No. 1	No. 2
Standapol® ES-1	30.00	60.00
Ceraphyl 60	2.00	2.00
Foamole A	1.00	1.00
Perfume	0.20	0.20
Kathon CG	0.05	0.05
Deionized Water	66.75	36.75

In suitable vessel weigh components in order listed. Completely solubilize perfume in surfactants before adding water, mix until uniform, and package.

Fill: 95% Concentrate
5% Propellant A-46
Precision Valve 2X.020 inverted, Foam Spout #02-1512

Dye-In Shampoo

Water	66.90
Deriphat 151C	20.00
Incronam 30	10.00
Incromide LR	1.00
Isobutyl Benzoate	0.50
Jaguar® C-14-S	0.30
Crotein Q	0.20
Germaben® II	1.00
Disodium EDTA	0.10
Dye	q.s.
Fragrance	q.s.

Reserve 10% of water to dissolve dye. Heat the rest of the water to 70 C and add **Jaguar® C-14-S, Crotein Q, Deriphat 151C**, isobutyl benzoate, **Incronam 30**, and **Incromide LR**. Add dye solution. Cool to 40–45 C, add fragrance and preservative.

Hair Conditioner

Formula No. 1

(Lotion)

A	Cetyl Alcohol	0.90
	Stearyl Alcohol	0.60
	Brij® 721	1.50
	Stearalkonium Chloride	0.25
B	**Arlatone PQ 225**	2.50
	Water	94.25

Heat Part A and Part B to 60 C. Add Part B to A with stirring. Cool by stirring to 40 C and package.

Physical Properties:

Typical Visc. — 2800 cps

Stability — 3 months at 25 C, 40 C, and 50 C

No. 2

(Cream)

Water Phase:	
Quatrisoft™ Polymer LM-200	1.0
Water	94.0
Oil Phase:	
Promulgen® D	4.5
Glyceryl Monostearate (neutral)	0.5
Perfume and Preservative	q.s.

Add **Quatrisoft™ Polymer LM-200** to water at room temperature with good mixing. When thoroughly dispersed, heat to 75 C. Add oil phase at 75 C to water phase at 75 C with mixing. Cool while mixing to room temperature.

No. 3

A	Glyceryl Stearate (pure)	4.00
	Foamine 1880	0.60
	Bernel Ester DOM	2.00
	Hetester ISS	2.00
	Cetyl Alcohol	2.00
	Parapel HC	2.50
	Arosurf TA-100	1.00
B	Deionized Water	83.05
	Cellosize® QP 30,000 Hi	0.50
	Parapel LAM-100	2.00
C	Lactic Acid (USP)*	0.25
D	**Kathon CG**	0.10

Heat Part A to 85 C and mix well. Heat Part B to 85 C and disperse. Add Part C to Part B and mix well. Continue to mix and slowly add Part A. Mix well without aeration at 80 C, then cool to 50 C and add Part D. Cool and mix to 30 C.

---

* The desired pH is 4.8–5.2 and can be adjusted in Part C.

No. 4

(Jojoba Bean Intensive Treatment)

Ameroxol® OE-10	1.5
Cellosize® QP 52000	1.0
CoJoba	0.1
Cycloton CTAC	2.5
Peptein 2000	0.3
Perfume, Dye, Preservative	q.s.
Water	94.6

Charge water into mixing vessel and heat to 60 C. In a separate container combine **Ameroxol® OE-10** and **CoJoba** and mix until completely dispersed. With smooth agitation, slowly add **Ameroxol® OE-10/ CoJoba** blend to heated water. Mix until uniform then cool with agitation to 50 C. Combine **Cycloton CTAC** and **Peptein 2000** and slowly add to water phase. With smooth agitation, lightly blend in **Cellosize® QP 52000** and mix until uniform. Cool with smooth agitation to 35 C and blend in perfume, dye, and preservative.

No. 5

(Balsam)

Deionized Water	80.0
Adogen 432-CG	2.5
Oleyl Alcohol	1.0
Cetyl Alcohol	2.1
Polawax®	1.0
Tween® 40	0.2
Peptein 2000	2.5
Balsam Fragrance	q.s.
Preservative, Coflor	q.s.
Water	q.s. to 100.0

Heat the water and **Adogen 432-CG** to 165 F. In another container heat the oleyl alcohol, cetyl alcohol, **Polawax®**, and **Tween® 40** to 165 F and add this slowly to the hot water phase while stirring. Cool the emulsion to 110–115 F and add the **Peptein 2000.** Adjust the pH to 4.5–5.5 with citric acid.

No. 6

(Premium, Balsam)

A	**Lexamine S-13**	3.00
	Lexemul CS-20	3.00
	Cetyl Alcohol	1.50
	Stearyl Alcohol	1.50
	Lexgard P	0.05
B	Hydroxypropyl Methylcellulose	0.80
	Propylene Glycol	2.00
	Lexein X250	1.00
	Lexgard M	0.15
	Balsam Fragrance	q.s.
	Phosphoric Acid	q.s. to pH 5
	Water	q.s. to 100.00

Disperse methylcellulose into 20% of the water at 80 C. Chill the rest of water, then add to the premix. Cool to 25 C and agitate until homogeneous. Reheat to 70 C. Add propylene glycol and **Lexgard M**. Adjust pH. Separately melt contents of Part A together. Add Part A to B with agitation. Cool to 55 C and add **Lexein X250**.

Physical Properties:

pH — 5.0

Visc. (@ 24 C) — 20,000 cps

Stability After 1 Month — Stable at room temperature, at 45 C, and passes 3 freeze/thaw cycles

No. 7

(Antidandruff)

Water	67.8
Alacsan 7-LUF	6.0
Siponic® C-20	1.5
Natrosol® 250HR (2.5%)	18.0
Sipoteric 1442	5.0
Glydant®	0.2

Zinc Omadine (48%)	1.0
FD&C Blue #1 (0.1%)	0.5
Fragrance	q.s.

Add **Alacsan** to water and heat to 65 C. Add **Siponic® C-20** with sufficient moderate agitation to produce a uniform dispersion. Discontinue heating. Add remaining ingrdients including previously prepared **Natrosol®** solution at room temperature singly with sufficient agitation to provide a uniform mixture before making next addition.

Note: The viscosity of the product may be increased by increasing the **Siponic® C-20** or **Natrosol®** concentration.

Physical Properties:

Appearance — Blue opaque liquid
Visc. (@ 25 C) — 3000 cps
pH — 5.6
Solids — 6.86%

Chlorine Neutralizing Conditioner

Water	86.30
Polymer JR-30M	0.70
Alacsan 7LUF	6.00
Siponic® C-20	1.50
PEG 6000 Distearate	0.25
Sipoteric 1442	5.00
Glydant®	0.20
Fragrance, Color	q.s.

Heat water to 60 C and dissolve **Polymer JR** with moderate agitation. Add **Alacsan** and heat to 65 C with continued agitation. Add **Siponic® C-20** and continue agitation to obtain a uniform dispersion. Add PEG 6000 distearate and continue agitation. After a uniform dispersion is obtained, discontinue heating. Add **Sipoteric 1442** and **Glydant®**. At 45–50 C add suitable fragrance and color.

Physical Properties:

Appearance — Emulsified white liquid
Visc. (@ 25 C) — 5500 cps
pH — 5.5
Solids — 6.85%

Perm Conditioner

Distilled Water	60.8
Alacsan 7LUF	6.0
Natrosol® 250HR (2.5%)	32.0
Collatein	1.0
Glydant®	0.2
Propylene Glycol	5.0
Fragrance, Color	q.s.

Dissolve **Alacsan 7LUF** in water with moderate agitation. Add **Collatein** with agitation. Add **Natrosol®** and **Glydant®**, singly with sufficient agitation to provide a uniform product before each addition. Dissolve appropriate fragrance in propylene glycol and add to batch with continued agitation. Add suitable color if desired.

Protein Conditioner

Formula No. 1

A	**Lexemul® EGMS**	0.50
	Cetyl Alcohol	0.50
	PEG 100 Stearate	0.50
	Lexgard P	0.05
B	Guar Hydroxypropyltrimonium Chloride	0.80
	Lexein QX3000	6.00
	Lexgard M	0.15
	Citric Acid	q.s. to pH 5
	Water	q.s. to 100.00

Heat water to 70 C. Dissolve **Lexgard M**. Disperse guar gum. Adjust pH. Separately melt contents of Part A together. Add Part A to B. Then add **Lexein QX3000**. Cool and fill.

Physical Properties:

Appearance–Opaque to semitranslucent
pH — 5.0
Visc. (@ 24 C) — 900 cps
Stability After 1 Month — Stable at room temperature, at 45 C, and passes 3 freeze/thaw cycles

No. 2

Mackpro NLP	3.0
Mackamide AME-100	2.5
Mackalene 426	2.0
DMDM Hydantoin	0.4
Natrosol® 250HHR	0.8
Lactic Acid	q.s. to pH 5.0
Water, Dye, Fragrance	q.s. to 100.0

Completely disperse **Natrosol® 250HHR** in water. Slowly add components and heat to 40 C. When product is clear, add fragrance and dye. Cool to room temperature.

Deep Conditioner

Formula No. 1

Emulsifying Wax (NF)	6.0
Propylene Glycol	3.0
Mackernium SDC-85	3.0
Mackester IDO	3.0
Glyceryl Monostearate	2.0
Mackalene 426	2.0
Mackpro NLP	1.0
Mackamide AME-100	1.0
DMDM Hydantoin	q.s.
Deionized Water, Fragrance	q.s. to 100.0

Melt first five components and heat to 150 F. Heat water, **Mackpro NLP,** and **Mackam AE-100** to 150 F. Slowly add water to the oil phase and blend for 30 min. Add preservative and DMDM Hydantoin at 110 F. Cool and fill.

No. 2

(Cream)

A	**Lexemul® 515**	6.00
	Lexamine S-13	2.50
	Lexol DIA	4.50

	Cetyl Alcohol	2.50
	PEG 400 Distearate	1.50
	Lexgard P	0.05
B	Glycerin	2.00
	Lexein X250	1.00
	Sodium Chloride	0.20
	Lexgard M	0.15
	Panthenol	0.05
	Citric Acid	q.s. to pH 5.0
	Water	q.s. to 100.00

Heat water to 70 C; add **Lexgard M** with mixing. Melt ingredients of Part A together. Add Part A to B with agitation, cool to 55 C, and add **Lexein X250** and panthenol.

Physical Properties:

pH — 5.0
Visc. (@ 24 C) — 2000 cps
Stability After 1 Month — Stable at room temperature, at 45 C, and passes 3 freeze/thaw cycles

No. 3

(High Activity)

Novol	1.00
Crodacol C-95	2.50
Incroquat SDQ-85	1.80
Croquat L	1.00
Kessco GMS (acid stable)	3.00
Carnation Mineral Oil	4.00
Glycerin	3.00
Germaben® II	1.00
Deionized Water	82.70

Heat the oil phase to 75 C. Heat the aqueous phase to 80 C. Add the water to the oil phase slowly. Continue to stir and fill off at 30 C.

Physical Properties:

pH — 4.6

Blow-Dry Conditioner

(Pump Type)

Alacsan 7LUF	0.6
Gafquat® 734	2.0
Liposorb L20	0.2
Perfume Oil	0.1
SD-40 Ethanol (anhyd.)	3.0
Distilled Water	94.1

Dissolve **Gafquat® 734** in water at room temperature. Add **Alacsan 7LUF**. In separate vessel, mix perfume, alcohol, and **Liposorb L20**. Add second mixture to first and blend thoroughly.

Mild Conditioner

(Salt Free)

Mackalene 426	8.0
Mackol 1618	4.0
DMDM Hydantoin	q.s.
Dye, Fragrance	q.s.
Water	q.s. to 100.0

Add **Mackalene 426** and **Mackol 1618** to water. Heat to 65–70 C and blend until homogeneous. Cool to 50 C. Add DMDM hydantoin, dye, and fragrance. Cool and fill.

Leave-On Conditioner

Formula No. 1

(Spray)

Mackpro NLP	1.0
Mackalene 426	3.0
DMDM Hydantoin	q.s.
Water, Dye, Fragrance	q.s. to 100.0

Add components to water. Heat to 40 C and blend until clear.

No. 2

(Clear)

Mackalene 426	6.0
Natrosol® 250 HHR	1.0
DMDM Hydantoin	q.s.
Deionized Water, Dye, Fragrance	q.s. to 100.0

Completely dispense **Natrosol®** in water. Add **Mackalene 426** and blend until clear. Heat to 40 C and add remaining components.

Comb-Out and Conditioner Spray

Glycerin	18.0
Propylene Glycol	18.0
PEG 75 Lanolin	0.7
Mackam CAP	0.3
Mackanate DC-30	0.2
Disodium EDTA	0.1
DMDM Hydantoin	q.s.
Deionized Water, Fragrance	q.s. to 100.0

Add components to water and heat to 40 C. Blend until clear. Adjust pH to 4.5–5.5 with citric acid.

Foam-On/Leave-On Hair Conditioner

Celquat® L-200 (2% aq. sol'n.)	50.00
Alcohol SD-40	30.00
Ceraphyl 65	0.75
Ceraphyl 85	0.50
Cetyl Alcohol	0.50
Deionized Water	18.25
Preservative	q.s.

In a suitable vessel weigh ingredients in order written and heat to 70 C with agitation. Keeping vessel tightly covered, mix until uniform, then start cooling. Continue cooling with agitation to 25–28 C and package.

Hair Dew

Oil Phase:	
Polawax®	0.20
Incroquat DBM90	0.50
Cosmowax J	0.30
Water Phase:	
Deionized Water	80.70
Incromate SDL	3.00
Methyl Paraben	0.30
Alcohol Phase:	
Alcohol SDA-40	15.00

Heat oil and water phase separately to 75 C then add water to oil under good agitation. Cool with mixing to 45 C then add alcohol. Cool with mixing to room temperature.

Note: Hair Dew is a no-rinse conditioner that is lightly sprayed onto wet or dry hair for instant conditioning without leaving the hair weighed down.

Physical Properties:

pH — 3.9

Creme-Rinse Conditioner

Formula No. 1

(High Viscosity)

A	**Carsoquat SDQ-85**	1.900
	Brij® 52	2.500
	Barlene® 18S	1.500
	Glyceryl Monostearate	0.500
	Stearyl Alcohol	0.100
	Phenoxyethanol	0.300
B	Citric Acid	1.750
	Sodium Chloride	0.200
	Deionized Water	91.215

C	Sodium Hydroxide (10% sol'n.)	0.035

Heat all components in Part A to approximately 60–65 C with mixing. With continuous agitation, heat Part B to approximately 65–70 C. Add Part A slowly to Part B. When blend is uniform, discontinue heating and add sodium hydroxide. Continue mixing the batch until temperature has cooled below approximately 35 C. Make up any water, dye, and fragrance lost in processing.

Note: The choice and amount of fragrance may alter the viscosity.

Physical Properties:

pH — 3.2 ± 0.1

Visc. (Brookfield RVT, #5 spindle @ 10 rpm, 25 C) — approx. 10,500 cps

No. 2

("Super" Active)

A	**Incroquat Behenyl TMS**	6.50
	Incromate SDL	5.00
B	**Incromectant AMEA-70**	7.00
	Crosilk Liquid	3.00
	Germaben® II	1.00
	Water	77.50

Heat Part B with protein and humectant to 75–80 C. Add the **Incroquat.** When uniform add the **Incromate SDL.** Continue stirring and fill at 25–30 C.

No. 3

(Opacified)

Carsoquat CT-429	3.50
Cetyl Alcohol	3.20
Citric Acid (anhydrous)	0.01
Water	93.29

No. 4

A	Cetyl Alcohol	1.5
	Stearyl Alcohol	1.5
	Parapel HC	2.5
	Arosurf TA-100	1.0
B	Deionized Water	92.9
	Cellosize® QP 30,000 Hi	0.5
C	**Kathon CG**	0.1

Heat Part A to 85 C. Heat Part B to 85 C and disperse well. Add Part A to Part B. Mix well at 85 C, then begin cooling. Continue mixing (avoid aeration) and at 50 C, add Part C. Mix until cool.

Note: Put on, rinse off type. Use $^1/_2$ to 1 oz.

No. 5

Deionized Water	67.00
Varisoft SDC	7.00
Glycol Stearate	1.60
Cetyl Alcohol	1.00
Water	16.00
Potassium Chloride	0.20
Peptein 2000	2.00
Perfume, Preservative, Color	q.s.
Water	q.s. to 100.00

Heat the water and **Varisoft SDC** to 165 F. Add the glycol stearate and cetyl alcohol and stir until dispersion is complete. Cool to 140 F and add the potassium chloride dissolved in water. Cool to 110 F and add the **Peptein 2000**, perfume, and color. A suitable preservative should be used.

No. 6

Distilled Water	46.8
Alacsan 7LUF	6.0
Siponic® C-20	2.0

Natrosol® 250HR (2.5%)	40.0
Sipoteric 1442	5.0
Glydant®	0.2
Fragrance, Color	q.s.

Add **Alacsan** to water and heat to 70 C with moderate agitation. Add **Siponic®** and stir until a uniform dispersion is obtained. Begin cooling to 50 C. Add **Natrosol®** solution, **Sipoteric**, and **Glydant®** with continued agitation. Add suitable fragrance and color.

Physical Properties:

pH — 6.4

Visc. — 7700 cps

No. 7

(Lotion)

Carsoquat SDQ-85	1.2
Cetyl Alcohol	1.2
Lonzest SMO	1.0
Stearyl Alcohol	0.8
Lonzest SMO-20	0.4
Water	q.s.
Preservatives, Perfume, Color	q.s.

In a large vessel heat water to 80 C. In small vessel combine **Carsoquat SDQ-85**, cetyl alcohol, **Lonzest SMO-20**, stearyl alcohol, and **Lonzest SMO**. Heat to 75 C with slight agitation, and while stirring, add components of small vessel to water in a large vessel. Begin cooling after $^1/_2$ hour.

No. 8

(Clear)

Carsoquat CT-429	5.00
Natrosol® 250 HHR	0.40
Polymer JR-400	0.10
Polyol 7000	5.00

Tetrasodium EDTA (38%)	0.30
Citric Acid	0.03
Water	89.17

No. 9

(Clear)

Lexein QX3000	6.00
PEG 150 Distearate	2.0 or q.s. to desired visc.
Lexgard M	0.15
Lexgard P	0.05
Water	q.s. to 100.00

Heat water to 70 C. Add **Lexgard** preservatives and **Lexein QX3000**. Slowly disperse PEG 150 distearate and mix until dissolved. Fill.

Note: The product can be easily opacified or pearlized by the addition of selected pigments.

Physical Properties:

pH — 6.2
Visc. (@ 24 C) — 850 cps
Stability After 1 Month — Stable at room temperature, at 45 C, and passes 3 freeze/thaw cycles

No. 10

(Clear)

Distilled Water	60.80
Alacsan 7LUF	6.00
Natrosol® 250 HR (2.5%)	32.00
Lexein X-350	1.00
Glydant®	0.20
Fragrance, Color	q.s.

Dissolve **Alacsan** in water using moderate agitation. Add previously prepared **Natrosol®** solution and stir until a satisfactory solution is obtained. Add **Glydant®**. Add suitable fragrance and color.

Note: A solubilized fragrance is recommended for this product.

Physical Properties:

pH — 6.0
Visc. — 900 cps

No. 11

(Aerosol Foam-On)

Lexein QX3000	6.00
Lexaine CSB-50	3.00
Laureth-23	2.00
Propylene Glycol	10.00
SD Alcohol 40	20.00
Water	q.s. to 100.00

Combine ingredients and mix until clear. Fill into a lined aerosol container with an 85:15 concentrate:propellant ratio.

Note: A rich foam with good spreadability is produced. To use, apply to wet hair, let sit 1 min, comb and rinse. This product will condition and lubricate the hair.

Conditioning Mousse

	Formula No. 1	No. 2
	(Rinse Out)	
Cremophor A-25	1.0	1.0
Luviquat FC-905	4.0	4.0
Luvitol EHO	1.0	—
Luvitol HP	—	2.0
Cetyl Stearyl Alcohol	4.0	4.0
Preservative	0.5	0.5
Fragrance	0.2	0.2
Water	89.3	88.3
Aerosol:		
Concentrate	90	90
Propane/Butane (40/60)	10	10

Package: Coated aluminum can and aluminum valve

No. 3

Distilled Water	62.0
Gafquat® 755N	7.5
Emulphor® ON-870	0.5
SD Alcohol 40	15.0
Fragrance	q.s.
Propellant A-46	15.0

Heat water to 45 C. Dissolve **Gafquat® 755N** and **Emulphor®** in water separately, keeping water at 45 C. Add alcohol. Cool to 40 C, add fragrance. Cool to room temperature. Charge with **A-46**.

No. 4

Distilled Water	63.27
Copolymer 937	5.00
Emulphor® ON-870	0.50
SD Alcohol 40 (190P)	16.23
Fragrance	q.s.
Propellant A-46	15.00

1. Dissolve **Copolymer 937** in water and heat to 45 C.
2. Dissolve **Emulphor® ON-870** in alcohol and heat to 45 C.
3. Add #2 to #1 and agitate until uniform.
4. Add fragrance and cool.
5. Charge into container and pressurize with **A-46.**

No. 5

Croquat M	2.50
Lanexol AWS	2.50
Incromectant LMEA	4.00
Incronam 30	0.50
Crodamol PMP	0.50
Dow Corning® 193	3.40

Glycerin	12.00
Propylene Glycol	5.00
Tween® 20	1.50
Germaben® II	1.00
Fragrance DR-07	0.50
Deionized Water	64.60
Incromate SDL	2.00

Premix the **Crodamol PMP, Croquat M, Lanexol AWS,** and **Incromectant LMEA.** When completely uniform, add to remaining ingredients, and heat to 75–80 C. Allow to cool. Pressurize and charge.

Suggested Propellant:

Propellant A-46	10%
Concentrate	90%

No. 6

A	**Cerasynt IP**	4.00
	Ceraphyl 50S	2.00
	Ceraphyl 41	2.00
	Cetyl Alcohol	1.50
	Ceraphyl 85	2.00
B	**Celquat L-200** (2% aq. sol'n.)	50.00
	Ethanol SD-40 (anhydrous)	30.00
	Deionized Water	6.50
	Ceraphyl 65	2.00

Fill:

Concentrate	95%
Propellant A-46	5%
Foam Spout #02-1512, Precision Valve	

In suitable vessel weight Part A and heat to 75–80 C with agitation. In another vessel able to contain the entire batch, weigh Part B and heat to 75–80 C with agitation. When proper conditions are obtained in both vessels slowly add Part A to B with agitation. Mix for 10–15 min and start cooling. Cool to 25–28 C with continuous agitation and fill.

No. 7

Distilled Water	64.18
Gafquat® 755N	7.50
PVP/VA E-735	2.00
Emulphor® ON-870	0.50
SD Alcohol 40 (190P)	10.82
Perfume, Color, Preservative	q.s.
Propellant A-46	15.00

1. Dissolve **Garquat® 755N** in distilled water. Heat to 45 C.
2. Dissolve **PVP/VA E-735** and **Emulphor® ON-870** in alcohol. Heat to 45 C.
3. Add #2 to #1. Mix until uniform.
4. Cool to room temperature.
5. Add perfume, color, and preservative.
6. Charge into container and pressurize with **Propellant A-46.**

No. 8

(Light-Styling)

Gaffix™ VC-713	5.0
Emulphor® ON-870	0.5
Dow Corning® 929	0.1
Ethanol	10.0
Distilled Water	69.4
Fragrance	q.s.
Propellant A-46	15.0

Mix all ingredients with gentle heating. Charge into container and pressurize with **A-46.**

Shampoo-In Hair Oxidation Color

Formula No. 1

(1:1 Dilution)

To prepare 1000 ml:

Carefully weigh the required quantities of the intermediates and transfer to a solution consisting of:

Deionized Water (50 C)	256.0
Kelene 77	15.0
Sodium Sulfite (anhyd.)	10.0
Sodium Isoascorbate	1.5
Propylene Glycol (USP)	86.0

Stir for 10–15 min at 50–55 C until solution is complete. Continue stirring. Then add:

Lowenol S-216X	222.0
Lowenol C-243 (60 C)	80.0
Oleic Acid (5° titer)	86.0
Ammonia CP Concentrate	125.0

Continue stirring. Cool to 30–40 C and then add:

Isopropyl Alcohol (99%)	125.0
Fragrance	q.s.

No. 2

(1:2 Dilution)

To prepare 1000 ml:

Carefully weigh the required quantities of the intermediates and transfer to a solution consisting of:

Deionized Water (50 C)	140.0
Kelene 77	15.0
Sodium Sulfite (anhyd.)	10.0
Sodium Isoascorbate	1.5
Propylene Glycol (USP)	86.0

Stir for 10–15 min at 50–55 C until solution is complete. Continue stirring. Then add:

Lowenol S-216X	308.0
Lowenol C-243 (60 C)	80.0
Oleic Acid (5° titer)	86.0
Ammonia CP Concentrate	150.0

Continue stirring. Cool to 30–40 C and then add:

Isopropyl Alcohol (99%)	125.0
Fragrance	q.s.

No. 3

(Cream)

Lowenol C-279	100.00
Lowenol C-243	40.00
Propylene Glycol (USP)	43.00

Heat the above to 60–70 C or until clear brilliant. Add:

Required grams oxidation intermediates

2,3 Dihydroxynaphthalene	0.65
Isopropyl Alcohol (99%, anhyd.)	62.50

Stir (heating to 60–70 C) until all solids are dispersed or dissolved. Add:

Oleic Acid (5° titer)	43.00
Lowenol S-216X	110.00
Monoethanolamine	25.00

With constant stirring, add a solution consisting of:

Deionized Water	563.00
Sodium Sulfite	5.00
Kelene 77	7.50

Stir for 20 min or until clear brilliant. Fill tubes.

Note: To apply, dissolve 8.60 g urea peroxide in 100.00 g of the above shampoo-in cream oxidation hair color or add 100.00 g 20 volume peroxide to 100.00 g cream oxidation hair color.

Hairdressing and Highlighter

Mackester TD-88	25.0
Lanolin (anhyd.)	17.0
Paraffin Wax	10.0
Mackester IDO	10.0
Mineral Oil	33.3
Mackanate DOS-70PG	4.0
BHT	0.2
Fragrance	0.2

Melt all ingredients at 70 C. Mix slowly while cooling and add fragrance at 38 C. Mix in and fill into jars.

Styling Glaze

Gaffix™ VC-713	7.5
Copolymer 937	5.0
Ammonyx KP	0.5
Dow Corning® 929	0.1
Natrosol® 250 HHR	0.5
Distilled Water	86.4
Fragrance, Color, Preservative	q.s.

Vortex water rapidly and sift in **Natrosol®**. When **Natrosol®** is wetted out, add remaining ingredients.

Note: A leave-on lotion formulated to impart style, conditioning, and luster to hair.

Styling Mousse

Formula No. 1

A	Arlatone PQ 220	10.00
	Brij® 72	1.00
	Stearalkonium Chloride	0.25
	Dow Corning® 193	0.50
	Water	73.25
B	SD Alcohol 40	10.00
C	**Propellant A-46**	5.00

Heat Part A to 60 C with stirring until uniform. Cool to 40 C with stirring and add Part B. Pack in suitable aerosol containers and pressurize with Part C at room temperature.

No. 2

(Light)

Gaffix™ VC-713	5.0
Emulphor® ON-870	0.5
Dow Corning® 929	0.1
Ethanol	10.0
Distilled Water	69.4
Fragrance	q.s.
Propellant A-46	15.0

Mix all ingredients with gentle heating. Charge into container and pressurize with **A-46.**

No. 3

(Non-Aerosol)

Gaffix™ VC-713	7.5
Alipal® HF-433	5.0
Gafamide® CDD-518	0.5

Ethanol	10.0
Distilled Water	77.0
Fragrance	q.s.

Mix all ingredients at room temperature until uniform.

No. 4

(Extra Hold)

Gaffix™ VC-713	7.5
Gafquat® 755N	4.0
Emulphor® ON-870	0.5
Dow Corning® 929	0.1
Alcohol SD 40	10.0
Distilled Water	62.9
Propellant A-46	15.0
Fragrance	q.s.

Dissolve **Gafquat®** in water and agitate until dissolved. Add remaining ingredients and agitate until uniform. Fill with propellant.

No. 5

(Aerosol Foam with Protein)

Copolymer 937	14.0
Distilled Water	55.4
Emulphor® ON-870	0.5
Alcohol SD 40	14.9
Kera-Tein 1000 AS	0.2
Fragrance	q.s.
Propellant A-46	15.0

1. Dissolve **Copolymer 937** in distilled water and heat to 45 C.
2. Dissolve **Emulphor® ON-870** in alcohol SD 40 and heat to 45 C.
3. Add #2 to #1 and agitate until uniform.
4. Cool to 40 C. Add protein and fragrance.
5. Charge into container and pressurize with **A-46.**

No. 6

(Conditioning)

Gafquat® 755N	4.0
Gaffix™ VC-713	7.5
Emulphor® ON-870	0.5
Dow Corning® 929	0.1
Alcohol SDA 40	10.0
Distilled Water	62.9
Propellant A-46	15.0
Fragrance	q.s.

Dissolve **Gafquat®** in water and agitate until dissolved. Add remaining ingredients and agitate until uniform. Fill with propellant.

No. 7

(Wet-Look for Ethnic Hair)

Distilled Water	40.3
Glycerin	10.0
Propylene Glycol	1.0
Copolymer 937	15.0
Kera-Tein 1000 AS	0.2
Mineral Oil	2.5
Emulphor® ON-870	1.0
SD Alcohol 40	15.0
Fragrance	q.s.
Propellant A-46	15.0

Heat water to 45 C. Dissolve glycerin, propylene glycol, and **Copolymer 937** (mixing well after each addition). Warm alcohol to 45 C. Dissove **Kera-Tein**, mineral oil, and **Emulphor®** (mixing well after each addition). Cool to 40 C. Add fragrance, adjust pH if desired. Cool to room temperature. Charge with **A-46**.

Liquefying Brilliantine Cream

Penreco Mineral Oil #9	51.08
Petrolatum Alba	32.80
Rosswax 60-0254	9.10
Ross Ceresine 1160/7	7.00
Caroteen	0.02

Mix oil ingredients in a steam-jacketed kettle at 165 F. When mixed, cool to 140 F and package.

Pomade

Formula No. 1

(Super Soft)

Syncrowax HRS-C	7.50
Crill 6	4.00
Crodamol CAP	2.50
Crodamol PTIS	1.00
Volatile Silicone 7207	85.00

Heat the **Crill 6, Crodamol CAP**, and **Crodamol PTIS** to 65 C. Slowly disperse the **HRS-C** and heat to approximately 110 C. Cool to 70 C and add **Volatile Silicone.** Stir and fill off.

No. 2

(Vegetable Hair)

Syncrowax HGL-C	7.75
Fluilan	6.00
Novol	42.05
Cholesterol (USP)	0.50
Crotein HKP	0.50
Crotein CAA	0.50
Tocopherol	0.10
Castor Oil (USP)	41.50
DL-Panthenol	0.10
Super Hartolan	1.00

Heat castor oil, **Syncrowax**, cholesterol, and **Novol** to 115 C, until a clear solution is formed. Cool slowly to 80 C and then add the lanolin, tocopherol, and panthenol and efficiently disperse the **Crotein HKP** and CAA. Then fill into jars.

No. 3

White Petrolatum	90.0
Monamid® S	5.0
Ritawax	5.0

Mix all ingredients together, heat with slight agitation until homogeneous.

Styling Gel

Water	94.3
Hetsorb L-20	2.0
Jaguar® CMHP	2.0
Borax	0.5
Poly-MAPTAC	0.5
Kathon CG	0.5
Herbal #137-047	0.2

Dissolve **Borax** in water. Add **Poly-MAPTAC**. Dissolve fragrance in **Hetsorb L-20**; add to water. Add **Jaguar® CMHP**. Heat to 40 C, adjust pH to 5–6 with citric acid. Add **Kathon CG**.

Physical Properties:

Visc. (Brookfield, RV-7, 20 rpm) — 27,000 cps

Setting Gel

Gaffix™ VC-713	7.5
Emulphor® ON-870	1.0
Cheelox® BF-13	0.1
Distilled Water	90.2

Carbopol® 940	0.5
Sodium Hydroxide (20%)	0.7
Preservative, Color, Fragrance	q.s.

Create a vortex in the water and sift in **Carbopol®** at room temperature. Vortex until **Carbopol®** is wetted out. Add sodium hydroxide, **Gaffix, Cheelox®.** Warm **Emulphor®** and add as a liquid (cool slightly before adding). Add color, fragrance, and preservative.

Pump-Spray Hair Set

Formula No. 1

Water	93.80
Poly-MAPTAC	4.00
Ammonyx CETAC	0.70
Tween® 20	0.60
Glycerin	0.20
Fragrance	0.20
Germaben® II	0.50

Dissolve fragrance in **Tween® 20.** To water add fragrance-surfactant mixture and rest of ingredients.

Hair Set Lotion

Formula No. 1

Distilled Water	77.90
Ethanol	20.00
Jaguar® C-17	1.00
Germaben® II	1.00
Polysorbate 20	0.10
Fragrance	q.s.
Color	q.s.
Citric Acid (50% sol'n.)	q.s.

To the water, add the **Jaguar® C-17** and agitate sufficiently to disperse. Now add the ethanol and polysorbate 20. Adjust the pH to 6.5–7.0 with citric acid. Now add preservative, fragrance, and color.

No. 2

Distilled Water	77.90
Ethanol SDA-40 (anhyd.)	20.00
Jaguar® C-17	1.00
Germaben® II	1.00
Hetsorb L-20	0.10
Fragrance	q.s.
Color	q.s.
Citric Acid (50% sol'n.)	q.s.

To the water, add the **Jaguar® C-17** and agitate sufficiently to disperse. Now add the ethanol and **Hetsorb L-20**. Adjust the pH to 6.5–7.0 with citric acid. Now add preservative, fragrance, and color.

No. 3

Luviskol VA 55 E	5.0–6.0
Luviquat FC 905	0.2–0.5
Ethanol or Isopropyl Alcohol	35.0
Distilled Water	q.s. 100.0

No. 4

Luviskol VA 64	2.0–3.0
Luviquat FC 905	0.2–0.5
Ethanol or Isopropyl Alcohol	0–10.0
Distilled Water	q.s. 100.0

No. 5

Luviquat FC 550	5.0–6.0
Ethanol or Isopropyl Alcohol	0–30.0
Distilled Water	q.s. 100.0

No. 6

Luviquat FC 370	5.0–6.0
Luviquat FC 905	0.2
Ethanol or Isopropyl Alcohol	0–30.0
Distilled Water	q.s. 100.0

Quick-Breaking Hair Cream

A	**Lexemul® 515**	3.00
	Lexol IPM	12.00
	Mineral Oil	20.00
	Stearic Acid	2.40
	Lexgard P	0.05
B	Propylene Glycol	5.00
	Triethanolamine	1.20
	Lexgard M	0.15
	Water	q.s. to 100.00

Add ingredients of Part B to water at 70 C. Separately melt contents of Part A together. Slowly add Part A to B with agitation. Cool to room temperature and fill.

Physical Properties:

pH — 8.4

Visc. (@ 24 C) — 40,000 cps

Stability after 1 month — Stable at room temperature, at 45 C, and passes 3 freeze/thaw cycles

Hair Sheen

(Water-Based)

A	Glycerin	3.00
	Lexein SLK	0.50
	Lexein X250	1.00
	Lexgard P	0.05
	Latex Opacifier	q.s.

B	Propylene Glycol	9.00
	Lexgard M	0.15
C	Carbomer 941	0.20
	Triethanolamine	q.s. to pH 7
	Water	q.s. to 100.00

Heat water to 65 C. Slowly add Carbomer 941 and neutralize, then add contents of Parts A and B. Cool and fill.

Physical Properties:

pH — 7.0

Visc. (@ 24 C) — 2375 cps

Stability After 1 Month — Stable at room temperature, at 45 C, and passes 3 freeze/thaw cycles

Hairspray

Formula No. 1

(Soft Hold)

Gantrez ES-225	3.00
Dimethylstearamine	0.21
Dow Corning® 190	0.10
Ethanol SD-40 (anhyd.)	71.69
Fragrance	q.s.
Propellant A-46	25.00

Physical Properties:

Cloud Pt. — < –20 F

No. 2

(Soft Hold)

Gantrex ES-425	3.00
AMP	0.06
Dow Corning® 190	0.10
Ethanol SD-40 (anhyd.)	71.84

Fragrance	q.s.
Propellant A-46	25.00

Physical Properties:
Cloud Pt. — < –20 F

No. 3

(Regular)

Gantrez ES-225	5.00
Dimethylstearamine	0.37
Dioctyl Sebacate	0.10
Ethanol SD-40 (anhyd.)	69.53
Fragrance	q.s.
Propellant A-46	25.00

Physical Properties:
Cloud Pt. — < –20 F

No. 4

(Super Hold)

Gantrez ES-225	7.00
Dimethylstearamine	0.50
Ethanol SD-40 (anhyd.)	67.50
Fragrance	q.s.
Propellant A-46	25.00

Physical Properties:
Cloud Pt. — < –20 F

No. 5

(Aerosol)

Butyl Ester of PVM/MA Copolymer	2.00
Lexein A210	0.50
SD-40 Alcohol	97.50

Combine ingredients and blend until clear. Fill into an aerosol spray container with **Propellant A-46** at 75:25 concentrate:propellant ratio.

Physical Properties:

Freeze/thaw (3 cycles) — passes
45 C Storage (3 months) — passes

No. 6

(Blow Dry)

Deionized Water	96.45
Incroquat Behenyl TMS	0.25
Incromectant LMEA	0.10
Incromate SDL	0.50
Lanexol AWS	0.30
Gafquat® 755N	2.00
Glydant®	0.30
Glycerin	0.10

Dissolve the **Gafquat®** in the water phase and heat to 75–80 C. Add the **Incroquat Behenyl TMS** at 75 C. When completely dissolved add the remaining ingredients. Continue stirring and fill off. Adjust pH to 5.8 with triethanolamine.

No. 7

(Conditioning and Spiking)

Water	86.00
Glycerin	3.20
Acetamide MEA	2.00
Polyquart H	2.00
Poly-MAPTAC	5.50
Germaben® II	1.00
Fragrance	0.30

Combine ingredients as listed.

Hair Relaxer

Formula No. 1

(Creme or Gel)

Miranol® DM	10.0 – 30.0
Sodium Hydroxide (50%)	6.0 – 10.0
Mineral Oil	5.0 – 5.0
Cetyl Alcohol	1.0 – 1.0
Cerasynt IP	3.0 – 3.0
Water	75.0 – 51.0

Melt together all the ingredients except the caustic soda and water. Prepare a solution of the caustic in water and add to the melt.

Physical Properties:
Solids — 14.5–21.5%

No. 2

(Cream)

Evanol	6.50
Duponol WA Paste	0.02
Ammonium Thioglycolate (60%)	11.68
Ammonia (28%)	3.87
Perfume Rose	0.30
Water	q.s. 100.00

If you wish to use thioglycolic acid (80%) in place of ammonium thioglycolate (60%), use:

Thioglycolic Acid (80%)	8.75
Ammonia (28%)	8.40

Heat water to 75 C. Add **Evanol** and **Duponol.** Keep at 75 C until the **Evanol** and **Duponol** are melted and dispersed. Stir to room temperature; add perfume. Add ammonium thioglycolate to the cream. Add sufficient ammonia to the cream to bring it up to pH 9.0.

Note: Assay thioglycolic acid content before packaging. It should be 7.0%.

No. 3

A	Cetyl Alcohol	2.00
	Crodafos N10N	1.50
	Volpo S-2	0.50
	Mineral Oil	13.00
	Petrolatum	11.50
	Stearic Acid (triple pressed)	8.00
B	**Volpo S-10**	2.50
	Croquat M	1.00
	Propylene Glycol	2.00
	Trisodium HEDTA	1.00
	Germaben® II	1.00
	Water	42.87
	Ammonium Thioglycolate	9.00
	Ammonium Hydroxide	4.63

Heat Part A to 80 C. Heat Part B to 80–85 C. Add Part B to A under good mechanical agitation. Preadjust the ammonium thioglycolate with ammonium hydroxide. Cool the emulsion to 45 C then add the ammonium thioglycolate solution. Fill off.

		No. 4 *(Caustic)*	No. 5 *(Bisulfite)*
A	**Veegum® HS**	2	2
	Water	53	57
B	Cetyl Alcohol	15	10
	Petrolatum	5	5
	Carnation White Mineral Oil	8	5
	Arlacel 165	5	5
C	Sodium Hydroxide	2	—
	Sodium Bisulfite	—	2
	Water	10	10
	Ammonium Carbonate	—	4
D	Preservative	q.s.	q.s.

Slowly add **Veegum® HS** to the water, while agitating at maximum available shear. Continue mixing until smooth. Heat Part A to 80 C. Heat Part B to 70 C. Add Part B to A with rapid agitation, cooling to 35 C. Slowly dissolve Part C ingredients and add to Parts A/B. Mix until smooth and uniform.

Consistency: Thick cream.

Suggested Packaging: Jar or squeeze tube.

No. 6

Incroquat Behenyl TMS	4.00
Polawax®	5.50
Volpo S-2	0.50
Crodafos N10N	1.50
Klearol Mineral Oil	11.00
Protopet Petrolatum	8.00
Crodacol C-95	1.00
Volpo S-10	2.50
Propylene Glycol	2.00
Germaben® II	1.50
Deionized Water	37.50
Calcium Hydroxide	5.00
Deionized Water	20.00

Heat the oil phase to 80 C. Heat the aqueous phase to 85 C. Add the aqueous phase to the oils under good mechanical agitation. Premix the calcium hydroxide solution. Allow emulsion to cool to 45 C, then add the premixture.

Note: Hair straightening preparations of this type may be covered by U.S. Patent 4,304,244.

Physical Properties:

pH — 12.8

Activator:

Deionized Water	68.40
Methyl Paraben	0.20
Germall® 115	0.20
Keltrol® F	0.20
Guanidine Carbonate	31.00

Dissolve the **Keltrol® F** and the water with stirring and heat to 70–75 C. Cool to 45 C and add the guanidine carbonate.

The relaxer is activated by adding 1 part of the activator with 4 parts of the cream.

No. 7

(Mousse)

Croquat M	1.10
Aminofoam C	3.00
Germaben® II	1.00
Deionized Water	81.00
Fragrance Q9721	0.40
Ammonium Thioglycolate (60%)	9.00
Ammonium Hydroxide	3.70

Heat the aqueous phase to 75 C. Allow to cool to 45 C, then add the fragrance. Mix thoroughly. Add premix solution of thioglycolate and ammonium hydroxide. Fill off. Charge.

Suggested Propellant:

Propellant A-31	10%
Base	90%

	No. 8 *(Average, Mousse)*	No. 9 *(Coarse, Mousse)*
Crodacol C-95	2.00	2.00
Crodafos N10N	1.50	1.50
Volpo S-2	0.50	0.50
Carnation Mineral Oil	13.00	13.00
Protopet	11.50	11.50
Stearic Acid (triple-pressed)	2.00	2.00
Propylene Glycol	2.00	2.00
Volpo S-10	2.50	2.50
Hamp-Ol Crystals	0.75	0.75
Germaben® II	1.00	1.00

Deionized Water	49.35	46.05
Croquat M	1.00	1.00
Ammonium Thioglycoate (60%)	9.00	11.50
Ammonium Hydroxide (29%)	2.90	3.70
Fragrance Q9721	1.00	1.00

Heat the oil phase to 80 C. Heat the aqueous phase to 80–85 C. Add the water to the oils under good mechanical agitation. Preadjust the ammonium thioglycolate with amonium hydroxide. Cool the emulsion to 49 C, then add the ammonium thioglycolate solution. Fill off. Charge.

Suggested Propellant:

Propellant A31	10%
Base	90%

Physical Properties:

pH — 9.5 ± 0.2

No. 10

Polawax®	7.50
Crodacol C-95	1.00
Protopet	4.00
Carnation Mineral Oil	15.00
Volpo S-2	0.50
Crodafos N10N	1.50
Propylene Glycol	2.00
Volpo S-10	2.50
Deionized Water	57.20
Sodium Hydroxide (25%)	6.80
Germaben® II	1.00
Croquat M	1.00

Heat the oil phase to 80 C and the aqueous phase to 85 C. Add the aqueous phase to the oils slowly under good mechanical agitation. Continue stirring and cool to 40–42 C, then incorporate the sodium hydroxide solution slowly.

Curl Activator

	Formula No. 1	No. 2
Glycerin	5.00	35.00
Propylene Glycol	35.00	5.00
Ceraphyl 140-A	5.00	5.00
Ceraphyl 41	5.00	5.00
Ceraphyl 65	2.00	2.00
Dow Corning® 193	2.00	2.00
Myrj 52S	1.50	1.50
Cetyl Alcohol	0.50	0.50
Glydant®	0.50	0.50
Deionized Water	43.25	43.25
Perfume M82-200	0.25	0.25

In suitable vessel weigh all ingredients and heat to 80–85 C with agitation. Mix until uniform, start cooling to 50 C and add perfume. Mix until uniform, continue cooling to 25–28 C and package.

Fill:	Concentrate	92%
	Propellant A-46	8%

No. 3

(Soft Gel)

Crodacol C-95	1.50
Super Refined Coconut Oil	2.00
Crodamol MM	1.50
Crodamol CAP	0.50
SF-96 Silicone (200 cps)	0.50
Incrocas 30	4.50
Incropearl NI Liquid	2.00
Dow Corning® 190	0.50
Crosilk 10,000	0.50
Crosilk Liquid	1.00
Crodafos SG	0.30
Incromectant AMEA 70	2.00
Incroquat SDQ-25	2.25

Propylene Glycol	8.00
Germaben® II	1.50
Deionized Water	71.15
Fragrance R994-R995	0.30
10% Triethanolamine (99%)	q.s. pH 5.0 ± 0.1

Heat the oil phase to 80 C. Add proteins and preservatives to aqueous phase and heat. When completely uniform, add the remaining ingredients and heat to 85 C. Add the aqueous phase to oils. Cool. Adjust pH to 5.0 ± 0.1 with 10% TEA solution.

Note: The **Super Refined Coconut Oil** can be replaced with **Super Refined Mink** or **Sesame Oil.**

No. 4

(Gel Microemulsion)

Crodafos N3N	7.48
Volpo 5	10.88
Mineral Oil	16.05
Incromectant AMEA-70	0.20
Sorbitol	15.00
Hexylene Glycol	5.00
Polymer JR-30M	0.10
Germaben® II	1.00
Water	44.29

Wave Set

Formula No. 1

(Mousse)

Water	55.8
Celquat® L200	8.0
Siponic® E-5	1.0
Ethyl Alcohol SD-40	20.0
Fragrance	0.2
Propellant A-46	15.0

Heat water to 60–65 C and dissolve **Celquat®** and **Siponic® E-5.** Dissolve fragrance in alcohol SD-40 and add to cooled water solution. Fill container and charge with **Propellant A-46.**

No. 2

(Pearlescent)

Croquat M	1.00
Crodapearl NI Liquid	1.00
Incronam 30	1.50
Cellosize® QP-4400	0.30
Hamp-ene Na_4	0.50
Germall® II	0.20
Triton® X-100	0.75
Propylene Glycol	4.00
Deionized Water	81.95
Ammonium Thioglycolate (60%)	6.00
Ammonium Hydroxide (29%)	2.40
Fragrance Q9721	0.40

Dissolve the gum in the water phase and heat to 85 C. When completely hydrated, add the preservative and **Hamp-ene Na_4**. Next add the protein. Then add the remaining ingredients. Cool to 49 C, add the premix solution of thio and hydroxide.

Physical Properties:
pH — 9.1

Wave Lotion

Formula	No. 1 *(Cold)*	No.2 *(Cold, Tinted)*
Water	69.06	72.55
Ammonium Thioglycolate (60%)	11.35	9.00
Hampex-80	0.20	0.20
Ammonia (28% aqueous)	4.14	3.00

Fragrance	0.15	0.15
Water	8.00	8.00
Emulsifier K-700	1.00	1.00
Sulfuric Acid	*	*
Ammonium (28% aqueous)	*	*
Water	*	*

Physical Properties:

pH	9.2–9.4	9.2–9.4
Free Ammonia	1.06–1.26 g NH_3/100 ml	0.74–0.94 g NH_3/100 ml

No. 3

(Cold, for Normal Hair)

Water	88.80
Thioglycolic Acid	5.65
Ammonia (28%)	2.05
Poly-MAPTAC	3.00
Fragrance	0.50

Adjust pH to 9.25–9.30.

	No. 4	No. 5
	(Self Timing)	
	(Regular)	*(Tinted)*
Water	50.00	55.00
Ammonium Thioglycolate (60%)	15.33	10.73
Diammonium Dithiodiglycolate (40%)	12.01	12.01
Hampex-80	0.25	0.25
Ammonia (28% aqueous)	2.86	2.14
Brij® 35	0.70	0.70
Fragrance	0.15	0.15
Water	10.00	10.00
Emulsifier K-700	1.00	1.00

*As needed

Sulfuric Acid	*	*
Ammonia (28% aqueous)	*	*
Water	q.s.	q.s.

Physical Properties:

pH	8.8–9.2	8.8–9.2
Free Ammonia	0.70–0.90 g NH_3/100 ml	0.50–0.70 g NH_3/100 ml
Thioglycolic Acid (%)	9.2 ± 0.1	6.44 ± 0.1

	No. 6	No. 7
	(Exothermic)	
	(Regular)	*(Tinted)*
Water	60.00	60.00
Ammonium Thioglycolate (60%)	27.60	21.97
Hampex-80	0.28	0.28
Ammonia (28% aqueous)	3.43	2.79
Brij® 35	0.78	0.78
Fragrance	0.18	0.18
Water	10.00	10.00
Emulsifier K-700	1.10	1.10
Sulfuric Acid	*	*
Ammonia (28% aqueous)	*	*
Water	q.s.	q.s.

Physical Properties:

pH	8.8–9.2	8.8–9.2
Free Ammonia	0.86–1.06 g	0.68–0.88 g
NH_3/100 ml	NH_3/100 ml	NH_3/100 ml
Fill Weight	95.0–98.1 g	94.0–97.1 g
Thioglycolic Acid (%)	16.56 ± 0.1	13.18 ± 0.1

*as needed to adjust pH

No. 8

(Acid)

A	Glycerol Monothioglycolate (80%)	
B	**Hampex 80**	0.25
	Emulsifier K-700	2.00
	Brij® 35	0.60
	Ammonia (28%)	*
	Phosphate Buffer	q.s.
	Water	q.s. to 100.00

Since glycerol monothioglycolate is not stable hydrolytically, Parts A and B are mixed just prior to use. The ratio of Parts A to B is 20–25 g of Part A to 60–80 ml of Part B with a resultant pH of 6.8–7.2.

No. 9

(Permanent)

A	**Veegum® HS**	1
	Water	69
B	Cetyl Alcohol	1
	Carnation White Mineral Oil	5
	Amerchol L-101	5
	Solulan® 98	2
	Arlacel 165	3
C	Ammonium Thioglycolate (60%)	12
	Ammonium Hydroxide	2
D	Preservative	q.s.

Slowly add **Veegum® HS** to the water, while agitating at maximum available shear. Continue mixing until smooth. Heat Part A to 80 C. Heat Part B to 75 C. Add Part B to A and mix until 35 C. Add Part C and mix until smooth and uniform.

---

*q.s. pH of mixture of Part A with Part B is 6.8–7.2

Consistency: Viscous lotion.
Suggested Packaging: Opaque plastic bottle.

Conditioning Perm

A	Cetyl Alcohol	2.00
	Crodafos N10N	1.50
	Volpo S-2	0.50
	Mineral Oil	13.00
	Petrolatum	11.50
	Stearic Acid	8.00
B	**Volpo S-10**	2.50
	Croquat M	1.00
	Propylene Glycol	2.00
	Trisodium HEDTA	0.50
	Germaben® II	1.00
	Water	42.87
	Ammonium Thioglycolate	9.00
	Ammonium Hydroxide	4.63

Heat the oil phase to 80 C. Heat the water phase to 80–85 C. Add the water to the oils under good mechanical agitation. Preadjust the ammonium thioglycolate with ammonium hydroxide. Cool the emulsion to 45 C, then add the ammonium thioglycolate solution. Fill off.

Leg Shave Cream

(Tube Dispensed)

A	**SF18** (350)	3.5
	Stearic Acid	3.5
	Cetyl Alcohol	1.0
B	**Carbopol® 934**	0.2
	Triethanolamine	1.0
	Propylene Glycol	7.5
	Veegum® (5% sol'n.)	20.0
	Sterile Water	63.3

C	Preservative	q.s.
	Perfume	q.s.

Heat Parts A and B to 70 C (158 F). Add Part A slowly to Part B with high-speed stirring. Continue slow mixing until temperature drops to 30 C (86 F). Blend in Part C.

Application: Apply a thin uniform film to prewetted skin and shave.

Depilatory

Formula No. 1

(Cream)

Evanol	6.50
Calcium Thioglycolate	5.40
Calcium Hydroxide	7.00
Duponol WA Paste	0.02
Sodium Silicate O	3.43
Perfume	as desired
Water	q.s.

Heat the water to 75 C. With stirring add the **Duponol, Sodium Silicate O**, and **Evanol**. Continue stirring at 75 C until the **Duponol, Sodium Silicate O**, and **Evanol** are dissolved. Remove the heat and continue stirring to room temperature. Add the calcium hydroxide and perfume. Finally, add the calcium thioglycolate until uniform. Assay for thioglycolic acid content of 2.7%. Package.

No. 2

(Lotion)

A	**Veegum® HS**	1
	Water	74
B	Propylene Glycol	5
C	**Amerchol L-101®**	5
	Solulan® 98	2
	Arlacel® 165	2

D	Calcium Thioglycolate	5
	Calcium Hydroxide	6
	Preservative	q.s.

Slowly add **Veegum® HS** to the water, while agitating at maximum available shear. Continue mixing until smooth. Add Part B to A and heat to 75 C. Heat Part C to 70 C. Add Part C to A/B with stirring. Cool to 45 C. Add calcium thioglycolate and stir. Add the calcium hydroxide and stir until cool and uniform.

Consistency Viscous lotion.

Suggested Packaging: Squeeze bottle or tube.

Preshave

Formula No. 1

Zinc Phenol Sulfonate	1.8
Ethyl Alcohol (denatured)	40.0
Menthol	0.1
Camphor	0.1
Water	58.0
Fragrance	q.s.

No. 2

Alcohol	50.3
Aluminum Chlorhydrate	2.0
Vegetol® Ivy MCF 775 (water soluble)	5.0
Perfume	0.5
Demineralized Water	42.2

All the components are mixed at room temperature.

No. 3

(Talc Stick)

A	**Veegum®**	1.9
	Water	q.s.
B	Zinc Stearate	4.7
	Magnesium Carbonate (light)	1.9
	Talc	91.5
C	Preservative	q.s.

Slowly add **Veegum®** to the water, while agitating at maximum available shear. Continue mixing until smooth. Blend Part B. Add Part A to B and work into a smooth paste. Add more water as desired. Pack in stick mold and allow to dry.

Suggested Packaging: Push or twist-up dispenser.

Brushless Shaving Cream

A	Deionized Water	78.193
	Glycerin	5.000
	Triethanolamine	0.857
B	**Methocel® 40-100**	0.100
C	Stearic Acid	10.000
	Stearyl Alcohol	0.500
	Acetylated Lanolin Alcohol	1.500
	Petrolatum (white)	1.500
	Glyceryl Stearate (SE)	1.500
D	Deionized Water	0.500
	Dowicil® 200	0.200
E	Perfume Oil (almond)	0.150

Mix Part A at room temperature for 10 min. Then add Part B and mix until completely dissolved. **Methocel®** will dissolve rapidly in the alkaline solution. Begin heating the solution, and in a separate vessel, combine and heat Part C. When both mixtures are between 75 and 80 C, combine with rapid agitation. When batch cools to below 45 C, add dissolved Part D. Add Part E when batch is below 40 C.

Shaving Gel

(Emulsifier-Free0

A	**Veegum®**	5.0
	Water	82.0
B	**Amerchol L-101®**	2.5
	Acetulan®	0.5
	Dow Corning® 200 Fluid (350 cs)	10.0
C	Preservative	q.s.

Slowly add **Veegum®** to the water, while agitating at maximum available shear. Continue mixing until smooth. Continue stirring, add Part B and mix until uniform.

Suggested Packaging: Jar or squeeze tube.

Shaving Foam

(Aerosol)

A	Myristic Acid	2.0
	Stearic Acid	4.0
	Isostearyl Isostearate	2.0
B	Coconut Fatty Acids	1.0
	Glycerin	1.0
	Polysorbate 20	6.0
	TEA-Lauryl Sulfate	1.0
	Sodium Borate	0.1
	Triethanolamine	2.5
	Demineralized Water	76.9
C	**Vegetol® Marigold** (water soluble)	3.0
	Perfume	0.5

Parts A and B are heated to 80 C. Part B is added to Part A while slow stirring. At about 50 C, **Vegetol®** and perfume are added.

The preparation is packaged in aerosol container with about 10% of propellant.

Shaving Cream

(for Ladies)

A	**Veegum®**	3
	Water	84
B	Glycerin	2
	Sorbitol (70%)	3
	Triton® X-100	3
C	**Myrj 45**	5
D	Preservative	q.s.

Slowly add **Veegum®** to the water, while agitating at maximum available shear. Continue mixing until smooth. Add Part B to A and heat to 70 C. Heat Part C to 75 C. Add Part C to Parts A/B and mix until smooth and uniform.

Consistency: Soft cream.

Suggested Packaging: Jar or squeeze tube.

After Shave

Formula No. 1

A	Alcohol	35.6
	Perfume	1.0
B	PPG-20 Methyl Glucose Ether	3.0
	Vegetol® Marigold MCF 774 (water soluble)	5.0
	Allantoin	0.1
	Demineralized Water	55.3

Part B is added to Part A at room temperature.

No. 2

("Proteinized" Skin Protector Cream, Nonionic O/W Emulsion Vehicle)

A	Stearyl Alcohol	4.0
	Cetyl Alcohol	3.0
	Sorbitan Stearate	2.0
	Mineral Oil (light)	4.0
	Mink Oil	1.0
	Squalane	1.0
	Octyl Dimethyl PABA	2.0
	Caprylic/Capric Triglyercide	2.0
B	Propylene Glycol	3.5
	Polysorbate 60	3.0
	Allantoin	0.2
	Citric Acid	0.1
	Preservatives	q.s.
	Deionized Water	to 100.0
C	**Sollagen**	5.0
	Firmenich 112.238*	3.0

Heat Parts A and B to 75 C with agitation and mix until homogeneous. Add part A to B. Continue agitation and cool to 35 C prior to adding Part C. Let stand overnight to promote viscosity buildup.

No. 3

(Lotion)

A	**Veegum®**	1.0
	Water	80.4
B	**Sipon® LSB**	3.4
	Glycerin	2.0
	Allantoin	0.1

* *Note:* This fragrance suggestion is an interesting spicy, woody, oriental blend with a crisp Amaretto topnote

C	**Carbowax® 400**	4.0
	Isopropyl Myristate	1.0
	Acetulan®	2.0
	Lecithin	1.0
D	SD Alcohol 40	5.0
	Menthol	0.1
E	Preservative	q.s.

Slowly add **Veegum®** to the water, while agitating at maximum available shear. Continue mixing until smooth. Add Part B to A and heat to 70 C. Heat Part C to 75 C. Add Part C to Parts A/B with mixing to 40 C. Add Part D and mix until cool and uniform.

Suggested Packaging: Bottle.

No. 4

(Lotion)

A	**Veegum®**	1.5
	Water	39.1
	Dow Corning® FG-10 Antifoam	0.2
B	**Dow Corning® 200 Fluid** (350 cs)	3.0
	Tween® 60	3.0
	Amerchol L-101®	1.5
C	**Klucel® GF**	1.5
	SD Alcohol 40	30.0
	Menthol (USP)	0.2
	Water	20.0
D	Preservative	q.s.

Slowly add **Veegum®** to the water, while agitating at maximum available shear. Continue mixing until smooth. Add antifoam. Blend Part B ingredients well. Add Part B to A while mixing. Prepare Part C by first dissolving menthol in alcohol and water. Slowly add **Klucel® GF**, mixing until dissolved. Add Part C to A/B and mix until emulsion is uniform.

Suggested Packaging: Squeeze bottle.

No. 5

(Skin Conditioner)

A	**Lanette O**	0.50
	Cetiol 1414-E	3.00
B	Water	68.45
	Carbopol® 934 (2% sol'n.)	15.00
	Glycerin	2.00
	Dowicil® 200	0.10
C	Triethanolamine (99%)	0.45
D	SD 40 Alcohol	10.00
	Fragrance	0.50

Heat Part A to 75–80 C. Heat Part B to 75–80 C. Add Part B to A under agitation. Add Part C. Remove heat. In a separate vessel dissolve fragrance in ethyl alcohol and add to batch at 45 C. Continue mixing until product reaches room temperature. Fill off.

No. 6

(Splash)

A	Water	90.4
	Carbopol® 1342	0.2
	Glycerin	2.5
	Hydrogenated Starch Hydrosylate	2.5
	DMDM Hydantoin	0.3
B	Mineral Oil	2.0
	Isopropyl Palmitate	2.0
	Fragrance	q.s.
C	Triethanolamine	0.1

Disperse **Carbopol® 1342** into water. Add remaining ingredients in Part A under mild agitation. Combine Part B; add to Part A with rapid agitation. Add Part C to the emulsion.

No. 7

(Soothing Gel)

A	Carbomer 941	0.25
	Water	63.05
B	**Quatrisoft™ Polymer LM-200**	0.25
	Water	9.75
C	SD Alcohol 40	15.00
D	Triethanolamine (99%)	2.50
	Water	9.20
	Perfume	q.s.

Prepare Part A and Part B separately with good agitation at room temperature until clear and uniform. If necessary, heat Part B after initial dispersing of **Quatrisoft™ Polymer LM-200**. Cool to room temperature. Add Part B to Part C. Add Part D to Part A, avoiding air entrapment. Add Parts B/C to A/D and mix until clear and uniform.

No. 8

(Gel)

A	Alcohol	35.60
	Carbomer 940	0.50
B	**Collagène 5015** (alcohol soluble)	1.00
	Demineralized Water	1.00
C	Demineralized Water	56.95
	Glycerin	2.00
	Allantoin	0.10
D	Perfume	0.50
	Solvating Agent γ 2428	1.00
E	Triethanolamine (50% aq. sol'n.)	1.35

Disperse Carbomer 940 in alcohol and add Part B while slow stirring. In this blend, add Part C and Part D. Add quickly the triethanolamine solution with sufficient stirring.

No. 9

(Alcoholic Emulsion)

A	**Tefose® 2561**	3.5
	Propylene Glycol Dipelargonate	3.0
B	Demineralized Water	68.5
	Kathon CG	0.1
	Glycerin	3.0
	Carbomer 941	0.2
C	Triethanolamine	0.2
	Demineralized Water	1.0
D	Alcohol	17.0
E	Soluble Collagen	1.0
	Demineralized Water	2.0
F	Perfume	0.5

Parts A and B are heated to 75 C. Add Part B to A. Cool while slow stirring. At about 70 C, add Part C, at about 25 C, add alcohol and Part E. Add perfume as required. Homogenize if necessary.

No. 10

(Emulsion)

A	PEG-20 Stearate	2.50
	PEG-8 Stearate	2.50
	Octyldodecyl Myristate	3.00
	Amerchol L101®	2.00

B	Demineralized Water	77.85
	Glycerin	2.00
	Allantoin	0.10
	Kathon CG	0.10
	Carbomer 941	0.30
C	Demineralized Water	5.00
	Sodium Hydroxide (10% aq. sol'n.)	1.00
	Triethanolamine	0.15
D	Soluble Collagen	1.00
	Demineralized Water	2.00
E	Perfume	0.50

Disperse Carbomer at room temperature. Parts A and B are heated to 80 C. Part B is added to Part A while slow stirring. At about 70 C, Part C is fastly added. At about 30 C, Part D is added. Add perfume as required. Homogenize if necessary.

Suntan Oil

	Formula No. 1	No. 2	No. 3	No. 4
SF1204	70	45	40	40
SS4267	5	5	5	5
Blandol	—	45	40	35
Fluilan	—	—	5	5
Almond Oil	—	—	—	5
Sunflower Oil	—	—	5	5
Escalol® 507	5	5	5	5
SDA40	20	—	—	—
Preservative	q.s.	q.s.	q.s.	q.s.
Perfume	q.s.	q.s.	q.s.	q.s.

Blend together all ingredients except for **SS4267**. Add the **SS4267** last and stir to a clear solution.

Note: Substitution of organic oils with silicones reduces oiliness and staining.

These formulas approximate SPF 6.

No. 5

(Greaseless)

Escalol® 507	5.00
Ceraphyl 368	47.50
Ceraphyl 41	10.00
Carnation Mineral Oil 65/75	27.50
Standamul® G-16	10.00

Weigh all ingredients in a vessel and mix until uniform.

Note: The expected SPF for this formulation should be 4–6. This value is only *approximated* and should be established by an outside laboratory.

No. 6

Drakeol 5	67.0
Isopropyl Palmitate	28.8
Lanolin (liquid)	2.5
Sun Screen	1.2
Perfume	0.5

Blend all the ingredients at room temperature and package.

Note: Spreads evenly and does not leave a greasy after feel on the skin.

No. 7

Coconut Oil	10
Dow Silicone 344	16
Isopropyl Myristate	13
Mineral Oil #7	42
Acetulan®	8
Glucam® P-20	4
Ross Jojoba Oil	2
Escalol® 507	5
Perfume	q.s.

Mix all of the above ingredients in a stainless steel vessel, run through a filter, and package.

Suntan Cream

Formula No. 1

A	Lanolin Alcohol Extract	5.00
	Acetylated Lanolin	5.00
	Ethoxylated Lanolin Alcohol (water soluble)	1.00
	Sun Screen	2.00
	Penreco Snow	5.00
B	Thickener	0.75
	Water	75.25
	Sodium Hydroxide (10% aq.)	2.25
	Ethoxylated Fatty Amine	3.75

Disperse the thickener into water using high speed agitation until the mixture is free of lumps. Heat the water/thickener blend to 70 C then mix in the sodium hydroxide solution and the ethoxylated fatty amine. Separately, mix the components of Part A together and heat them to 75 C. Blend Part A into B and continue to mix until cooled to ambient temperature. Perfume may be added at 40 C.

Note: A white opaque cream leaves a silky feeling on the skin.

No. 2

(Water-Resistant)

A	**Veegum®**	2.0
	Water	69.5
	Glycerin	3.5
B	**Amerscreen P**	2.0
	Arlacel® 20	3.5
	Tween® 20	4.5
	Dow Corning® 556 Fluid	5.0
	Stearic Acid (triple-pressed)	5.0
	Kessco Glycerol Monostearate SE	5.0

C Preservative	q.s.

Slowly add **Veegum®** to the water, while agitating at maximum available shear. Continue mixing until smooth. Add the glycerin and mix until uniform. Heat Part A to 70 C. Heat Part B to 75 C. Add Part B to A, mixing until cool.

Suggested Packaging: Opaque tube.

	No. 3	No. 4	No. 5
	(SPF 4)	*(SPF 6)*	*(SPF 8)*
A **SF97** (50)	0.5	0.5	0.5
SF1202	8.5	8.5	8.5
SS4267	5.0	5.0	5.0
Sunflower Oil	4.5	4.5	4.5
Cetyl Alcohol	0.8	0.8	0.8
Stearic Acid	4.0	4.0	4.0
Escalol® 507	3.0	5.0	4.0
Uvinul® M40	—	—	2.0
B Distilled Water	67.5	65.5	64.5
Triethanolamine	1.2	1.2	1.2
Carbopol® (1% sol'n.)	5.0	5.0	5.0
C Perfume	q.s.	q.s.	q.s.
Preservative	q.s.	q.s.	q.s.

Blend Part A ingredients, adding **SS4267** last. Parts A and B are heated separately to 70 C (158 F). Part B is slowly added to Part A with thorough mixing. The emulsion is cooled to less than 45 C (113 F) and Part C is blended in.

Note: This recipe should result in a smooth, light cream suitable for dispensing from a squeeze tube or bottle.

No. 6

(Self-Tanning)

A	**Tefose® 1500**	10.00
	Stearic Acid	3.50
	Lanolin Oil	1.50
	Labrafil® M 2130 CS	3.00
	Mineral Oil	6.00
	Fondix® G BIS	1.00
	Menthyl Salicylate	3.00
B	Demineralized Water	67.85
	Dihydroxy Acetone	4.00
C	Perfume	0.15

Parts A and B are heated to 70 C. Part B is slowly added to Part A. Cool while stirring. Add perfume as required. Homogenize if necessary.

		No. 7	No. 8
		(Water-Resistant)	
A	**Dow Corning Fluid® 200** (100 cs)	15.00	—
	Escalol® 507	7.00	7.00
	Syntase 62	3.00	3.00
	Stearic Acid (triple-pressed)	5.00	5.00
	Cetyl Alcohol	1.00	1.00
	Ceraphyl 368	—	15.00
B	Deionized Water	65.00	65.00
	Cellosize® QP 30,000	0.50	0.50
	Triethanolamine (85%)	2.50	2.50
C	**Germaben® II**	1.00	1.00

In a suitable vessel, weigh Part A and heat to 80–85 C with agitation. In another vessel able to contain the entire batch, weigh water and heat to 80 C. Completely dissolve **Cellosize®** with agitation and add triethanolamine. When proper conditions are obtained in both phases, slowly add Part A to Part B and allow to mix for 10–15 min (homo-mill or high-shear lighting). Start cooling, continuing agitation to 40 C and add Part C. Continue mixing while cooling to 25–28 C and package.

Suntan Cream/Lotions

	Formula No. 1	No. 2	No. 3 *(Anionic)*	No. 4	No. 5
A **Cerasynt MN**	10.00	10.00	10.00	10.00	10.00
Emulsynt 1055	3.00	3.00	3.00	3.00	3.00
Ceraphyl 368	6.00	6.00	6.00	6.00	6.00
Escalol® 507	3.00	5.00	4.00	5.00	7.00
Uvinul® M40	—	—	2.00	2.50	3.00
Stearic Acid (triple-pressed)	4.00	4.00	4.00	4.00	4.00
Cetyl Alcohol	0.75	0.75	0.75	0.75	0.75
Dow Corning® 200 Fluid (100 cs)	0.75	0.75	0.75	0.75	0.75
Propyl Paraben	0.10	0.10	0.10	0.10	0.10
B Deionized Water	64.95	62.95	61.95	60.45	57.95
Triethanolamine (88%)	2.00	2.00	2.00	2.00	2.00
Glycerin	5.00	5.00	5.00	5.00	5.00
Glydant®	0.25	0.25	0.25	0.25	0.25
Methyl Paraben	0.20	0.20	0.20	0.20	0.20
Approximate SPF*	≈ 4	≈ 6	≈ 8	≈ 12	≈ 15

Heat Parts A and B with mixing to 80 C. Add Part A slowly to Part B (which is mixing) at 80 C and continue to mix for 10 min. Cool and mix to 45 C and add the perfume. Continue cooling to 25–28 C (avoid aeration).

* Actual values to be ascertained by laboratory testing

Suntan Lotion

Formula No. 1

A	**Carbopol® 940** (2% aq. sol'n.)	25.0
	Deionized Water	42.9
	Propylene Glycol	4.4
	Methyl Paraben	0.2
B	Mineral Oil	15.0
	Myristyl Myristate	2.0
	Cetyl Alcohol	1.0
	Lanolin Alcohol	1.0
	Stearic Acid	2.4
	Propyl Paraben	0.1
	Octyl Methoxycinnamate	5.0
C	Triethanolamine (99%)	1.0
D	Fragrance	q.s.

Combine the **Carbopol®** resin mucilage and water using low shear agitation. Add the propylene glycol and methyl paraben and heat to 70 C. Weigh all the ingredients in Part B into a separate vessel and heat to 70 C. Add Part B to Part A using moderate shear agitation. Add Part C using moderate shear agitation. Cool, with stirring, to 40 C. Add Part D, stirring well until room temperature is reached.

No. 2

(Extra Rich)

A	**Veegum®**	1.0
	Water	86.0
	Propylene Glycol	3.0
B	Stearic Acid (triple-pressed)	2.0
	Dow Corning® 556 Fluid	1.0
	Cetyl Alcohol	0.5
	Acetulan®	4.0
	Giv-Tan F	2.0
	Triethanolamine	0.5

C	Preservative	q.s.

Slowly add **Veegum®** to the water, while agitating at maximum available shear. Continue mixing until smooth. Add the propylene glycol and mix until uniform. Heat Part A to 70 C. Heat Part B to 75 C. Add Part B to A, mixing until cool and uniform.

Consistency: Flowable lotion.

Suggested Packaging: Opaque squeeze or pump bottle.

Note: The sunscreen is estimated to provide a SPF in the 2–4 range. This base would be suitble for use with other sunscreen agents.

Self-Tanning Lotion

A	**Tefose® 1500**	8.0
	Stearic Acid	0.8
	Mineral Oil	6.0
B	Demineralized Water	50.0
	Kathon CG	0.1
	Carbomer 941 (1% gel)	10.0
C	Demineralized Water	20.9
	Dihydroxyacetone	4.0
D	Perfume	0.2

Parts A and B are heated to 75 C. Part B is added to Part A. Cool while stirring. At about 30 C, add Part C. Add perfume as required. Homogenize if necessary.

Suntan Stick

(Water-in-Oil)

A	**Veegum®**	1.0
	Water	40.0
	Glycerin	3.0

B	Mineral Oil	12.5
	Acetulan®	5.0
	Dow Corning® 200 Fluid (350 cs)	5.0
	Petrolatum	4.0
	Paraffin	20.0
	Stearic Acid (triple-pressed)	4.0
	Giv-Tan F	2.0
	Pationic CSL	3.0
	Witcamide 511C	0.5
C	Preservative	q.s.

Slowly add **Veegum®** to the water, while agitating at maximum available shear. Continue mixing until smooth. Add glycerin and heat to 85–90 C. Heat Part B with stirring to 85–90 C. Add Part A to B and mix until smooth. Pour into molds and cool.

Suggested Packaging: Opaque push or twist action dispenser.

Note: The approximate SPF should be in the 2–4 range.

Sunscreen

Formula No. 1

Ross White Bleached Beeswax	20.2
Ross Pure Refined Candelilla Wax	16.1
Ross Pure Refined #1 Yellow Carnauba	4.0
Petrolatum	16.1
Isopropyl Myristate	10.0
Mineral Oil (60/70)	32.4
Ross Jojoba Oil	1.2
Amerscreen P	q.s.

Heat all ingredients in a steam-jacketed kettle to 170 F under agitation. When fully mixed cool to 145 F and package.

No. 2

A	Stearic Acid (triple-pressed)	4.00
	Escalol® 507	4.00
	Coconut Butter	3.00

	Lonzest 143-S	2.00
	Cerasynt 945	1.00
B	**Carbopol® 941** (2% sol'n.)	10.00
	Water	75.50
	Triethanolamine	0.50

Sunscreen Oil

Formula No. 1

A	Coconut Oil	10.0
	Labrafil® Isostearique	20.0
	Oily Vegetol® Marigold	15.0
	Butyl Methoxydibenzoylmethane	1.0
	Octyl Methoxycinnamate	3.0
B	Mineral Oil	50.2
	Antioxydant WL 774	0.3
	Perfume	0.5

Part A is heated to 40 C until butyl methoxydibenzoylmethane is dissolved. Then add the other components.

	No. 2	No. 3	No. 4	No. 5	No. 6
			(Super Refined)		
Super Refined Almond Oil	94.4	92.4	91.4	88.4	86.4
Super Sterol Ester	5.0	5.0	5.0	5.0	5.0
d,l Alpha Tocopherol	0.5	0.5	0.5	0.5	0.5
Escalol® 507	—	2.0	3.0	5.0	5.0
Uvinul® 539	—	—	—	1.0	3.0
Propyl Paraben	0.1	0.1	0.1	0.1	0.1
Approximate SPF value	0	2	4	8	15

Combine ingredients and mix. Slight heat may be used to aid processing.

Sunscreen Cream

Formula No. 1

Bentone® Gel MIO	10.0
Ritachol®	3.0
Glyceryl Stearate	5.0
Polysynlane	2.0
Isocetyl Stearate	10.0
Lexol PG 8-10	16.0
Escalol® 507	3.0
Stearic Acid	5.0
Propyl Paraben	0.1
Deionized Water	44.6
Triethanolamine	0.9
Methyl Paraben	0.1
Fragrance	0.3

No. 2

A	**Apifil®**	12.5
	Mineral Oil	9.0
	Coconut Oil	7.0
	Apricot Kernel Oil	4.0
	Petrolatum	7.5
	Labrafil® 2130 CS	5.0
	Oily Vegetol® Marigold	2.0
	Octyl Methoxycinnamate	3.0
B	Demineralized Water	39.7
	Sorbitol (70% sol'n.)	10.0
	Kathon CG	0.1
C	Perfume	0.2

Parts A and B are separately heated to 75 C. Part B is slowly added to Part A while stirring slowly for 5 min. Then cooling is started until 25 C. Add perfume as required.

No. 3

A	**Apifil®**	15.0
	Mineral Oil	22.5
	Apricot Kernel Oil	7.5
	Oily Vegetol® Marigold	3.0
	Octyl Methoxycinnamate	6.0
B	Demineralized Water	36.3
	Propylene Glycol	5.0
	Titanium Dioixde	4.0
	Kathon CG	0.1
C	Perfume	0.6

Parts A and B are heated to 75 C. Part B is added to Part A. Cool while stirring slowly. At about 25 C, add perfume. Homogenize if necessary.

Cocoa Butter Sunscreen Cream Base

A	Cocoa Butter	8
	Cetyl Alcohol	3
	Stearic Acid	2
	Ross Beeswax Substitute 628/5	5
	Glycerol Monostearate	9
	Arlacel® C	1
B	Water	69
	Borax	3

Melt Part A in a kettle at 185 F. Disperse ingredients of Part B at 185 F. Under agitation add Part B to A at 160 F. Cool to 130 F and pass through a homogenizer. Pour into containers.

Sunscreen Cream/Lotion

(Water-Resistant)

A	**Escalol® 507**	7.00
	Spectra-Sorb® UV-9	3.00
	Dow Corning® 200 Fluid	1.00
	Emulsynt GDL	1.00
	Ceraphyl 368	10.00
	Cetyl Alcohol	0.50
	Cerasynt 945	5.00
	Petrolatum	2.00
	Cerasynt 840	1.00
	Mineral Oil	10.00
B	**Carbopol® 934** (2% sol'n.)	10.00
	Propylene Glycol	5.00
C	Deionized Water	34.40
	PVP/K30	2.00
	Sodium Hydroxide (10% sol'n.)	0.80
D	**Carboset XL-40**	6.00
	Germaben® II	1.00
	Perfume	0.30

Combine Part A and heat to 80 C. Combine Part B and heat to 80 C. Disperse PVP in water. Add sodium hydroxide. Add PVP solution to Part B. Add Part A to water phase and maintain heat and mixing 10–15 min to emulsify. Cool to 40 C and add Part D in order listed. Continue mixing and cool to room temperature.

Sunscreen Lotion

Formula No. 1

A	**Veegum®**	0.60
	Rhodigel 23	0.15
	Water	65.45
	Unimulse CG	0.80
	Propylene Glycol	3.00

B	**Finsolv® TN**	8.00
	Uvinul® M-40	5.00
	Parsol® MCX	7.50
	Ritachol®	4.00
	Tween® 80	0.50
C	Zinc Oxide	5.00
D	Preservative	q.s.

Blend **Veegum®** and **Rhodigel**. Slowly add to the water, while agitating at maximum available shear. Continue mixing until smooth. Add remaining Part A ingredients slowly in order, mixing after each until uniform. Mix Part B well until the **Uvinul® M-40** dissolves. Add Part B to A and mix until smooth. Add Part C and mix until smooth and uniform.

Suggested Packaging: Opaque squeeze container.

No. 2

A	Water	76.5
	Stearamidopropyl PG-Dimonium Chloride Phosphate	3.0
	Sorbitol (70%)	5.0
B	Glycerol Stearate	3.0
	Dimethicone + Trimethylsiloxysilicate	3.0
	Myristyl Myristate	1.0
	Octyldimethyl PABA	5.0
	Oxybenzone	3.0
	Tocopheryl Acetate	0.5

Heat Part A to 65 C with stirring. Heat Part B to same temperature with stirring. Add Part A to B with agitation, and stir while cooling to 40 C. Add fragrance, coloring, or preservative as required and package while warm.

Physical Properties:

pH — 6.6

No. 3

A	Deionized Water	80.5
	Carbopol® 1342	0.2
	Sorbitol (70%)	2.0
B	Propylene Carbonate	0.5
	Isopropyl Myristate	5.0
C	Triethanolamine	0.2
	PEG-15 Cocamine	0.2
D	Octyl dimethyl PABA	1.5
	Benzophenone-3	2.0
	Benzyl Alcohol	7.5
E	Methyl Paraben	0.2
	Propyl Paraben	0.2
F	Fragrance	q.s.

Disperse **Carbopol® 1342** into water, then add sorbitol. Combine Part B. Add to Part A with rapid agitation. Add Part C to form the emulsion. Combine Part D until clear yellow solution results. Add Parts D and E to the emulsion. Add fragrance.

No. 4

A	**Tefose® 1500**	7.0
	Stearic Acid	1.0
	Oily Vegetol® Marigold	4.0
	Mineral Oil	3.0
	Octyl Methoxycinnamate	3.0
	Carotene (oily sol'n.)	0.1
	Lipocire DM	0.5
B	Demineralized Water	77.9
	Kathon CG	0.1
	Carbomer 941	0.1

C	Triethanolamine	0.1
	Demineralized Water	3.0
D	Perfume	0.2

Parts A and B are heated to 75 C. Part B is added to Part A. Cool while stirring slowly. At about 25 C, add part C. Add perfume as required. Homogenize if necessary.

	No. 5	No. 6
	(Foams)	
4-Aminobenzoic Acid	5.0	5.0
Brij® 721	2.0	2.0
Lauryl Alcohol	4.0	4.0
Fragrance	0.1	0.1
Propylene Glycol	2.0	2.0
Water	79.4	82.9
Dymel® 22	3.0	4.0
A-17 (*n*-Butane)	4.5	—
Properties:		
Vapor Pressure @ 70 F (psig)	48	64

Foam Appearance: Formula No. 5 is a stiff, expanding and pressure-sensitive foam, whereas Formulas No. 6 is a soft and creamy foam.

Sunscreen Gel

(Super Refined)

Super Refined Shark Liver Oil	86.95
Super Sterol Ester	5.00
Syncrowax HGL-C	2.00
Syncrowax ERL-C	1.00
BHA	0.05
Escalol® 507	5.00

Combine and heat with agitation to 80 C. When completely uniform, cool to 40 C and fill off.

Sunscreen Mousse

	Formula No. 1	No. 2	No. 3	No. 4
Lusantan-25	8.0	8.0	8.0	5.0
Cremophor A-25	1.0	1.0	0.5	1.0
Cetyl Alcohol	1.5	1.5	1.0	1.0
Miglyol	5.0	5.0	6.0	—
Mineral Oil	5.0	5.0	6.0	5.0
Eusolex 3573	2.0	2.0	2.0	—
Dimethicone	—	0.1	0.2	0.5
Stearic Acid	3.5	3.0	2.0	3.0
Triethanolamine	0.3	0.3	0.4	0.3
Luvitol EHO	—	—	—	5.0
Propylene Glycol	—	3.0	3.0	3.0
Luviquat FC-370	—	—	—	1.0
Preservative	0.2	0.2	0.2	0.2
Fragrance	0.2	0.2	0.2	0.2
Water	73.3	71.2	70.5	74.8
Aerosol				
Concentrate	90	90	90	95
Propane/Butane (40/60)	10	10	10	5

Package: Aluminum can and aluminum valve

No. 5

(Water-Resistant)

A	**Dow Corning® 200 Fluid** (100 cs)	15.00
	Escalol® 507	7.00
	Syntase 62	3.00
	Stearic Acid (triple-pressed)	5.00
	Cetyl Alcohol	1.00

B	Deionized Water	45.00
	Cellosize® QP 30,000	0.50
	Triethanolamine (88%)	2.50
	Alcohol SD 40	20.00
C	**Germaben® II**	1.00

In a suitable vessel weigh Part A and heat to 80 C with agitation. In another vessel with the capacity to contain the entire batch, weigh water and completely dissolve **Cellosize®** while heating to 80 C. At 80 C slowly add TEA/alcohol blend and mix until uniform, keeping vessel tightly covered. When proper conditions are obtained in both vessels, slowly add Part A to B and mix for 10 min. Start cooling, continuing agitation to 40 C and add Part C. Continue mixing while cooling to 25–28 C and package.

Sunscreen Stick

Syncrowax HGL-C	20.0
Syncrowax ERL-C	6.0
Super Refined Mink Oil	7.0
Stearic Acid (triple-pressed)	3.0
Crodamol PMP	32.5
Castor Oil	21.0
Cab-O-Sil® M-5	0.5
Escalol® 507	5.0
Uvinul® N539	3.0
Silicone L45 Fluid (200 cS+)	2.0

Combine ingredients and heat to 80 C with mixing, until uniform. Cool. Pour off at 58 C into antiperspirant/deodorant sized containers. Cool to room temperatures. Flame.

Physical Properties:

Melting Pt. (Drop Pt. Method) — 50 C

Penetration (102.5 g cone, no weights, 5 s @ 25 C) — 5.4 mm

Emollient Sunblock

Escalol® 506	2.500
Ceraphyl 50	5.000
Carbowax ®1500	5.000
Foamole A	2.500
FDC Brown	0.005
Ethanol (95%)	52.500
Water	32.495

Dissolve all ingredients except water in alcohol. Slowly with constant agitation, add water.

Note: This is a fast-drying preparation.

Sunscreen Moisture Gel

(All-Weather Sportsman for Men)

A	Deionized Water	68.8
	Natrosol® 250 HR	0.9
	Nutri-Tein BMV	2.0
	Polypro 15,000	0.5
	Lactamide MEA, Cocoyl Hydrolyzed Animal Collagen, and Glycerin	1.0
	Uvinul® MS-40	1.0
	Collagen Glyceride	1.0
	PABA	0.5
B	**Sollagen**	3.0
C	Alcohol SD-40	20.0
	Polysorbate 20	0.8
	Fragrance	0.5
D	Color to suit	q.s.

Heat water to 60–65 C and sprinkle the **Natrosol®** into the stirred water. Allow to fully dissolve to a clear solution. Stop heating and add ingredients listed in Part A. Cool to below 35 C and add Part B (**Sollagen**). Premix Part C and, finally, add Part D. Adjust pH to 6.5.

After-Sun Lotion

Formula No. 1

(With Jojoba Oil)

A	Mineral Oil (60/70)	6.2
	Modulan®	5.0
	Rosswax 63-0412	6.6
	Propylene Glycol	3.3
	Ross Jojoba Oil	1.7
B	Water	67.7
	Aloe Vera Liquid	3.3
	Triethanolamine	1.2

Melt Part A and Part B in separate vessels to 170 F under agitation. When temperature is reached, mix Part A to Part B and cool. Package in containers at below 120 F.

No. 2

A	**Tefose® 2561**	6.0
	Oily Vegetol® Marigold	5.0
	Carbomer 941 (1% gel)	10.0
	Demineralized Water	40.0
B	Perfume	0.2
	Kathon CG	0.1
	Demineralized Water	33.5
	Allantoin	0.2
	Soluble Collagen	2.0
	Glycerin	3.0

Parts A and B are heated to 75 C. Part B is added to Part A. Cool while stirring slowly. At about 25 C, add perfume. Homogenize if necessary.

No. 3

A	Water	17.70
	Carbopol® 941 (1% sol'n.)	50.00
	Dowicil® 200	0.10
B	**Lantox 55**	6.00
	Wecobee M	5.00
	Cetiol G-16S	8.00
	Cetiol LC	8.00
	Emulgade F	5.00
C	Sodium Hydroxide (50%)	q.s. to pH 6.5
D	Fragrance	0.20

Disperse **Carbopol®** in water. Mix and heat Part A to 60–65 C. Mix Part B and heat to 60–65 C. Add to Part A. Add Part C. Cool to 30 C and add Part D.

After-Sun Emollient

Oil Phase:	
Crodamol W	10.00
Polawax®	5.00
Super Refined Sesame Oil	3.00
Super Refined Coconut Oil	5.00
Escalol® 507	2.00
Water Phase:	
Deionized Water	72.60
Germaben® II	1.00
Carbomer 941	0.20
Veragel Lipoid 1:1	1.00
Triethanolamine (99%)	0.20

Charge a vessel with water and dust in the carbomer. Start heating to 75 C. When carbomer is completely dissolved add in **Germaben®**. Combine and heat oil phase to 75 C. Add triethanolamine to water. Add the water to oil phase and mix well. At 40 C add **Veragel** and cool to room temperature.

Toothpaste

Formula No. 1

(Flowable Gel)

A	**Veegum®**	1.00
	CMC 7MF	0.25
	Water	21.25
B	Sorbitol (70%)	12.50
	Glycerin	12.50
C	Dicalcium Phosphate Dihydrate	50.00
D	Flavor	1.00
E	**Maprofix® 563**	1.50
F	Preservative	q.s.

Blend **Veegum®** and **CMC**. Slowly add to the water, while agitating at maximum available shear. Continue mixing until smooth. Add Part B to A while mixing. Add parts C and D while mixing until smooth. Add Part E with slow mixing. Avoid incorporation of air.

Suggested Packaging: Plastic squeeze bottle.

Fluoride Toothpaste

		Formula No. 1 *(Thick Gel)*	No. 2 *(Thin Gel-Pump)*
A	**Veegum®**	1.50	0.75
	Water	24.50	19.10
B	Sorbitol (70%)	36.50	50.00

C	Glycerin	10.00	10.00
	CMC 7MF	—	1.00
D	**Syloid® 74**	24.00	13.00
	Syloid® 63	—	3.00
E	Flavor	1.20	1.00
F	Sodium Fluoride	0.25	0.25
	Saccharin	0.20	0.20
	Sodium Benzoate	0.20	0.20
	Maprofix® 563	1.50	1.50
	Citric Acid	0.15	—
G	Color	q.s.	q.s.

Slowly add **Veegum®** to the water, while agitating at maximum available shear. Continue mixing until smooth. Mix in Part B, followed by Part C (preblend CMC into glycerin for Formula No. 2). Mix until smooth. Add parts D, E, and F in order, mixing after each addition until smooth and uniform.

	No. 3	No. 4
	(Dicalcium Phosphate-Based)	
Dicalcium Phosphate Dihydrate	51.75	50.00
Water	23.55	21.30
Glycerin	20.00	12.00
Sorbitol (70%)	—	12.00
Sodium Lauryl Sulfate	2.00	2.00
Viscarin® TP 328	1.00	1.00
Fluoride MFP®	0.75	0.75
Sodium Benzoate	0.50	0.50
TSPP	0.25	0.25
Saccharin	0.20	0.20
Color and Flavor	to suit	to suit

Hot Process: Blend **Viscarin® TP 328** with saccharin, benzoate, and fluoride. Add to glycerin or glycerin/sorbitol mixture while at room temperature and with good agitation. Add water and heat to 150–160 F (65–70 C). Add sodium lauryl sulfate. Slow agitation to prevent foaming. Stop heating. Add abrasive while constantly mixing. Mix for 10–15 min and cool before adding volatile flavor oils. Mill and deaerate. Fill tubes at 85–105 F (35–40 C).

No. 5

(Silica-Based)

Water	33.95
Syloid® 63	21.00
Glycerin	18.00
Sorbitol (70%)	18.00
Syloid® 244	5.00
Sodium Lauryl Sulfate	2.00
Sodium Benzoate	0.50
Disodium Phosphate	0.40
Sodium Saccharin	0.20
Viscarin® TP 389	0.77
Sodium Fluoride	0.18
Flavor and Color	to suit

Hot Process: Add **Viscarin® TP 389** to glycerin or glycerin/sorbitol mixture while agitating. Add hot water to bring temperature up to 105 F (40 C). Add benzoate and fluoride after the carrageenan has hydrated. Add sodium lauryl sulfate and **Syloids®**. Finish process the same as in preceding formula.

Cold Process: Disperse **Viscarin® TP 389** in half of the glycerin before adding to the sorbitol and water. Mix in Ross mixer for 5 min at a setting of 3.5. Add **Syloids®** and mix at number 5 speed setting for 5 min. Disperse sodium fluoride, disodium phosphate, color, sodium saccharin. Add dispersion to mixer and agitate for 5 min at number 3.5 setting. Add sodium lauryl sulfate and continue mixing for 5 min. Add flavor oil and mix for 2 min. Deaerate before filling into tubes.

No. 6

(Silica-Based Gel)

Water	33.14
Syloid® 63	21.00
Glycerin	18.00
Sorbitol (79%)	18.00
Syloid® 244	5.00
Sodium Lauryl Sulfate	2.00
Viscarin® TP 389	0.80
Flavor Oil	0.80
Sodium Benzoate	0.50
Disodium Phosphate	0.40
Sodium Saccharin	0.16
Sodium Fluoride	0.18
Color	0.02

Cold Process: Disperse **Viscarin® TP 389** into half of the glycerin. Add to sorbitol and water in a Ross vacuum mixer and mix at medium speed for 2 min. Add **Syloids®** and mix for 2 min at medium speed. Mix for 15 min at high speed. Disperse remaining dry ingredients into the remaining glycerin before adding to mixer. Add and mix 5 min at medium speed. Add flavor oil and mix for 2 min. Vacuum deaerate for 10–15 min before filling into pump dispenser.

Tooth Cream

(Dicalcium Phosphate-Based)

Dicalcium Phosphate	52.00
Water	25.27
Glycerin	18.00
Sodium Lauryl Sulfate	1.50
Viscarin® TP 339	1.00
Flavor Oil	0.80
Fluoride MFP®	0.75
Sodium Benzoate	0.50
Trisodium Polyphosphate	0.25
Sodium Saccharin	0.20

Hot Process: Disperse the **Viscarin®** and salts (except for sodium lauryl sulfate) into the glycerin using a Ross mixer. Add water and heat to 150–160 F (65–70 C). Hold at 155 F (68 C) for 15 min while continuously agitating. Add dicalcium phosphate while mixing at low speed. Mix at high speed for 10 min. Add sodium lauryl sulfate and flavor oil. Mix for 2 min. Vacuum deaerate prior to filling pump dispenser.

Denture Cleaner Tablet

A	**Veegum® WG**	5
	Sodium Perborate	13
	Tetrasodium Pyrophosphate (anhydrous)	25
	Sodium Chloride	13
	Tartaric Acid	9
	Sodium Phosphate (dibasic)	12
	Citric Acid	7
	Sodium Bicarbonate	16
B	Isopropyl Alcohol	25
C	**Veegum® WG**	2

Granulate Parts A with B and pass through a No. 10 mesh screen. Dry and pass through a No. 16 mesh screen. Dry blend with Part C and compress.

Consistency: Tablet.

Suggested Packaging: Bottle or individual wrap.

Note: **Veegum® WG** is used as both a binder and disintegrant in this tablet. The **Veegum® WG** provides a hard tablet. It also aids disintegration of the tablet when water is added.

Chapter V
COATINGS

Green Primer

A	**Versamid 280-B75**	280.0
	Genamid 2000	20.0
	Mistron M139	500.0
	Titanium Dioxide R-960	100.0
	n-Butyl Alcohol	264.0
	Yellow Iron Oxide YO-3587	20.0
	Phthalo Blue BT417D	1.0
	Thixatrol ST	5.0
B	**Epoxy Resin EEW-180**	500.0
	Mistron M139	100.0
	Super Hi-Flash Naphtha	258.0
	Celite 289	150.0
	Thixatrol ST	5.0

Pigments are dispersed to give an N.S. fineness of grind reading of at least 4 on Part A and at least 2 on Part B. Parts A and B are thoroughly mixed 1:1 by volume and allowed to induct approximately 30 min prior to use.

Note: The above components must reach **Thixatrol ST** activating temperature.

Physical Properties:

Total Solids — 73%
P/B Ratio — 1.19/1
Wt./Gal — 10.7 lb
PVC — 33.2%
Drying Time (firm) — 6 h

Drying Time (hard) — overnight
Pot Life — 6 h
60° Gloss — 15
Visc. — 77–87 KU

Air-Dry Red Iron Oxide, Medium Oil Primer

Formula No. 1

(Solvent-Based, Alkyd)

	lb/100 gal
Pigment Grind:	
Medium Oil Alkyd Resin	176.0
Bentonite Clay Gel (12.5% active)	22.0
Soya Lecithin	4.0
Mineral Spirits	32.0
Butrol 22	50.0
Magnesium Silicate	165.5
Red Iron Oxide	275.0

Disperse on high-speed disperser to Hegman 4.

Letdown:	
Medium Oil Alkyd Resin	312.0
Cobalt Naphthenate (6%)	3.0
Manganese Naphthenate (6%)	2.0
Zirconium Drier (6%)	6.0
Antiskinning Agent	1.0
Mineral Spirits	68.0

No. 2

(Water-Reducible, Alkyd, with 0.5 lb/gal Inhibitor)

	lb/100 gal
Pigment Grind:	
Water	135.0
Busperse 39	6.0
Bubreak 448	3.0
Zinc Sulfate	2.0

Igepal®CTA-639		2.0
Butrol 22		50.0
Red Iron Oxide		100.0
Magnesium Silicate		200.0

Disperse on high-speed disperser to Hegman 4.

Letdown:

Water-Reducible Alkyd Resin	Premix before	240.0
Sec. Butanol	adding to	89.0
Ammonium Hydroxide (28%)	grind	10.0
Water		232.0
Cobalt Drier (6%)		2.0
Zirconium Drier (6%)		4.0

Red Iron Oxide Primer

A	**Epoxy Resin EEW-525** (75% in toluene)	227.3
	Beetle 216-8	10.4
	Red Iron Oxide	418.6
	Celite 266	68.8
	Emtal® 41	56.6
	Aluminum Stearate #907	3.4
	MIBK	80.4
	Ektasolve EB	9.0
	Toluene	32.7
B	**Versamine A52**	88.9
	Denatured Ethanol	21.2
	Toluene	90.4
	Ektasolve EB	9.0
	MIBK	80.3

Thoroughly mix Parts A and B prior to use. Allow 30–60 min induction time before application. If needed, reduce viscosity using MIBK, **Ektasolve EB**, toluene, and denatured alcohol (40/4.5/45/10.5 ratio).

Physical Properties:

Wt./Gal — 12.0 lb
Solids (wt.) — 68.2%
Solids (vol.) — 45.5%
Typical Tack-Free Time — 1.5 h
Typical Throughcure Time — 48 h

Chemical-Resistant Primer

(Rule 66 Exempt)

A	**Epon® 1001-T-75**	163.0
	Strontium Chromate 5006	107.0
	Titanium Dioxide (Type 111, Class A TT-P-442)	20.0
	Asbestine 625	49.0
	Diatomaceous Silica (Mil-S-15191)	26.0
	Methyl Isobutyl Ketone (TT-M-268)	18.0
	Xylene	27.0
	Isopropanol	23.0
	Methyl Ethyl Ketone	102.00
B	**Versamid 115-I73**	93.0
	Isopropanol	72.0
	Butanol	63.0
	Cellosolve	129.0

Thoroughly mix Parts A and B prior to use. Allow 15–30 min induction time before application. Reduce viscosity as needed after induction period.

Physical Properties:

Wt./Gal — 8.95 lb
Total Vehicle Solids (wt.) — 21.31%
Total Pigment Solids (wt.) — 22.64%
Solids (vol.) — 28.16%
P/B Ratio — 1.06:1.00
Coating Visc. (#4 Ford Cup) — 16 s
Typical Hard-Dry Time — 7–8 h

Rust-Inhibiting, Lacquer-Resisting Synthetic Primer

Formula No. 1

Zinc Chromate Y-539-D	58.60
Brown Iron Oxide B-2093F	316.44
Nytal® 300	210.96
Metasap-R	7.03
Aroplaz 3667-Z-80	175.80
Methyl Isobutyl Ketone	117.20

Disperse to 7 N.S.

Aroplaz 3667-Z-80	157.17
Lead (24%)	5.56
Cobalt (6%)	2.23
Exkin #2	1.34
Methyl Isobutyl Ketone	117.20

Physical Properties:

P/B Ratio (wt.) — 2.2/1.0
Total Solids (wt.) — 74%
Visc. (#4 Ford Cup) — approx. 30–35 s

Note: Meets most of performance requirements of Fed. Spec. TT-P-664C

No. 2

Zinc Chromate Y-539-D	56.53
Brown Iron Oxide B-2093F	305.24
Nytal® 300	203.49
Metasap-R	6.78
Aroplaz 3667-Z-80	169.58
Xylene	56.53
Methyl Isobutyl Ketone	56.53

Disperse to 7 N.S.

Aroplaz 3667-Z-80	151.60
Lead (24%)	5.36
Cobalt (6%)	2.15
Exkin #2	1.29
Xylene	73.48
Methyl Isobutyl Ketone	73.48

Physical Properties:
P/B Ratio — 2.2/1.0
Total Solids (wt.) — 72%
Visc. (#4 Ford Cup) — approx. 30 s

Note: Meets most of performance requirements of Fed. Spec. TT-P-664C

Corrosion-Inhibiting Maintenance Primer

A	**Kelpoxy 519**	259.91
	Halox CW-291	83.04
	Synthetic Yellow Iron Oxide YI0-2288D	186.85
	Sparmite	62.29
	Nytal® 300	80.97
	Kelecin F	3.11
	100 Type-Aromatic Solvent	28.03
	Xylene	134.95
	Methyl Isobutyl Ketone	32.54
	Isopropyl Alcohol	20.77
	Butoxy Ethanol	25.94
	Beetle 216-8	10.40

Pebble mill to 6 N.S. grind.

B	**Versamid 115**	61.77
	Isopropyl Alcohol	24.08
	Xylene	61.46

Physical Properties:
P/B Ratio: 1.7/1.0
Solids (wt.) — 59.9%
Equivalent Ratio — 1.0/1.0
Visc. (#2 Zahn Cup) — 30–40 s

Metal Maintenance Primer

(Water Reducible)

A	**Genamid 5701-H65**	198.17
	Water	258.25
	Halox CW-22	198.77
	Imsil 1240	44.58
B	**Genepoxy M200**	157.16
	Dowanol DPM	90.93
	Kroma Red 6097	167.49
	Asbestine 325	77.59
	Imsil® 1240	316.04
	Cellosolve Acetate	130.76

Primer for Polyethylene

	Formula No. 1	No. 2
CP-153-2 (25%)	20.0	10.0
Elvax 260	—	2.5
Toluene or Xylene	80.0	87.5

Aqueous Emulsion Primer for Polypropylene

A	**CP-343-3** (25%)	25.000
	Toluene	24.925
	Sulfonated Castor Oil Surfactant	0.025
	Texapon 2HC	0.050
B	Deionized Water	50.000

Primer for Polypropylene and Other Plastics and Metals

	Formula No. 1	No. 2	No. 3	No. 4
CP-343-1 (25%)	20.0	10.0	—	—
CP-343-1 (50%)	—	—	10.0	5.0
Elvax 260	—	2.5	—	2.5
Toluene	80.0	87.5	90.0	92.5

	No. 5	No. 6	No. 7	No. 8
CP-343-1 (100%)	5.0	2.5	—	—
CP-343-3 (25%)	—	—	20.0	10.0
Elvax 260	—	2.5	—	2.5
Toluene	95.0	95.0	80.0	87.5

Stir-In Additive in Coatings for Polypropylene and Other Plastics

	Formula No. 1	No. 2
CP-343-3 (25%)	10.0	—
CP-515-2 (40%)	—	7.5
Alkyd Urethane, Acrylic Enamel, or Lacquer	100.0	100.0

Note: Level of chlorinated polyolefin may be adjusted to give optimum properties.

Plastisol Primer

Ucar® Phenolic Resin BKS-2750	41.90
Dicyandiamide	1.31
Dimethylformamide	7.45
Cyclohexanone	21.90
Methyl Isobutyl ketone	16.48
Carbitol®	10.96

Add the solvents, except dimethylformamide, to **BKS-2750** solution. Dissolve the dicyandiamide in the dimethylformamide and add this solution to the diluted **BKS-2750** solution.

Physical Properties:

Solids — 26.28%
Sp. Gr. — 0.960
Wt./Gal — 8.01 lb
Visc. (Gardner) — As

Note: Apply by spray, dip, or roller coat. Maximum performance is obtained at a 0.3–0.7 mil film thickenss. Bake for 5–10 min at 400 F (204 C) or follow an equivalent schedule.

Medium-Oil Alkyd Primer

(Nontoxic)

Aroplaz 1445-M-50	252.80
M-34 Manganese Oxide	183.75
Busan 11-ML	105.00
Bentone® SD-1 Clay	9.45
Nytal® 400	126.00
10 Wollastokup® WC	105.00

Disperse to a $4^1/_2$ N.S.U. then add:

Aroplaz 1445-M-50	167.20
Mineral Spirits	157.50
Zirconium Drier (12%)	5.25
Cobalt Drier (12%)	2.10
Manganese Drier (6%)	1.05
Calcium Drier (4%)	3.15
Methyl Ethyl Ketoxime	1.05

Physical Properties:

Nonvolatiles (vol.) — 42.91%
PVC — 44.07%

Phenolic Alkyd Primer/Topcoat

(Black)

Aroplaz 1365-X-60	188.91
CKU 2266 X-50	137.02
Zinc Oxide XX631	12.75
Strontium Chromate J-1365	12.75
Lampblack	3.99
B-2093F Brown Iron Oxide	51.81
Bentone® SD-1 Clay	6.38
Nyad® 1250	278.98

Disperse on a high-speed disperser to 5 N.S.U. then add:

Toluene	305.29
Zirconium Drier (12%)	2.79
Cobalt Drier (12%)	0.60
Methyl Ethyl Ketoxime	0.80
Triethylamine	2.39

Physical Properties:

Nonvolatiles (vol.) — 35.30%
PVC — 48.77%

Long-Oil Alkyd Primer/Topcoat

(Inhibitive, One-Coat)

Aroplaz 1266-M-70	259.68
Rutile Titanium Dioxide	156.00
Bentone® SD-1 Clay	8.42
Halox CW-2230	208.00
10 Wollastokup® WC	151.01
Beaverwhite 325 Talc	52.00

Disperse to a 5 N.S.U. then add:

Aroplaz 1266-M-70	136.00
Mineral Spirits	169.00
Zirconium Drier (12%)	5.72

Cobalt Drier (12%)	1.30
Manganese Drier (6%)	1.56
Methyl Ethyl Ketoxime	1.04

Physical Properties:
Nonvolatiles (vol.) — 55.15%
PVC — 42.34%

Long-Oil Alkyd Primer

(Chromate-Based)

Aroplaz 1266-M-70	280.00
Zinc Oxide XX631	53.11
Strontium Chromate J-1365	53.11
M-34 Manganese Oxide	159.32
Bentone® SD-1 Clay	8.60
Beaverwhite 325 Talc	212.43
10 Wollastokup® WC	159.32

Disperse to a $4^1/_2$ N.S.U. then add:

Aroplaz 1226-M-70	102.40
Mineral Spirits	179.50
Zirconium Drier (12%)	5.69
Cobalt Drier (12%)	1.26
Manganese Drier (6%)	1.53
Methyl Ethyl Ketoxime	1.06

Physical Properties:
Nonvolatiles (vol.) — 54.16%
PVC — 43.22%

Alkyd Urea Baking Primer

(Nontoxic)

Aroplaz 1453-X-50	222.40
N-Butanol	17.47
Xylene	34.82

Bayferrox 140	126.01
Molywhite 101	198.24
Bentone® SD-1 Clay	11.52
10 Wollastokup® 10014	149.92

Disperse to a 5 N.S.U. then add:

Aroplaz 1453-X-50	207.40
Xylene	69.63
N-Butanol	34.82
Beetle 216-8	89.21
Catalyst 4040	2.85
Triethylamine	1.24

Bake for 15 min at 350 F.

Physical Properties:
Nonvolatiles (vol.) — 42.44%
PVC — 35.18%

Alkyd/Melamine Baking Primer

(High-Solids)

Aroplaz 6230-A6B-90	280.80
Arcosolve PMA Acetate	147.23
M-34 Manganese Oxide	254.16
Nalzin 2	149.00
Bentone® SD-1 Clay	7.89
10 Wollastokup® 10014	438.20

Disperse to a $4^1/_2$ N.S.U. then add:

Methyl Ethyl Ketone	42.07
N-Butanol	21.77
Beetle 216-8	22.35
Resimene 747	95.18
Catalyst 4040	1.75

Bake for 13 min at 400 F.

Physical Properties:

Nonvolatiles (vol.) — 67.09%

PVC — 43.27%

Phenolic/Phenoxy Baking Primer

Ucar® Phenolic Resin BKR-2620	9.01
Ucar® Phenoxy Resin PKHH	3.15
Iron Oxide (synthetic)	12.16
Barytes	13.51
Zinc Oxide	1.35
Methyl Ethyl Ketone	24.33
Cellosolve Acetate	24.33
Toluene	12.16

Dissolve **BKR-2620** and **PKHH** in the MEK/**Cellosolve Acetate**/ toluene solvent blend. Charge the resin solution plus pigments to a ball mill and grind 16–24 h.

Application: Apply by spray to a dry film thickness of approximately 1 mil. For optimum properties, use a minimum bake of 20 min at 375 F (191 C) or 15 min at 400 F (204 C).

High-Solids Baking Enamel

(White)

	Formula No. 1 *(15 Vol.% Hydrous Kaolin)*	No. 2 *(10 Vol.% Calcined Kaolin)*
High-Solids Polyester Resin	111	111
Titanium Dioxide	402	426
Hydrous Kaolin Extender Pigment	45	—
Calcined Kaolin Extender Pigment	—	30
Methyl Ethyl Ketone	46	46
Ethylene Glycol Monoethyl Ether Acetate	52	52

Disperse on a high-speed disperser for 10 min:

High-Solids Polyester Resin	411	411
Melamine Crosslinker	148	148
Mar and Slip Aid	2	2
Acid Catalyst	10	10

High-Sollds Baking Coating

Ucar® Phenolic Resin CK-1634	100
Tung Oil*	135
Xylene**	60

Dissolve **CK-1634** in solvent and add tung oil. Manganese drier (0.02% metal on oils) may be added to speed bake cure.

Application: Apply and bake for 10 min at 400 F (204 C).

Physical Properties:

Visc. (Gardner K) — 275 cP
Oil Length — 17 gal
Nonvolatiles — 79.7%
Solids (vol.) — 77%

Phenolic Pipe Baking Coating

Iron Oxide (pure)	29.95
Bentone® 27	1.51
Ucar® Phenolic Resin BKS-2600	41.02
Ethanol	18.40
Butanol	4.56
Butyl Cellosolve	4.56

Charge all ingredients to pebble mill and grind to a good dispersion.

* Alternate: Tung oil/alkali-refined linseed oil (60/40).

** Ethylbenzene/Cellosolve solvent/Rule 66-type mineral spirits (20/20/60).

Baking Schedule:

Maxmimum Performance —	1 h/275 F (135 C)
	2 h/350 F (177 C)
Minimum Bake —	20 min/350 F (177 C)

Physical Properties:

Wt./Gal — 10.53 lb
Pigment Vol. Ratio — 30%
Nonvolatiles — 53.5%
Visc. — 67 K.U.

Industrial Maintenance Enamel

(White)

A	**Kelpoxy 519**	195.75
	Titanox® 2160	183.62
	Methyl Isobutyl Ketone	227.78
	Beetle 216-8	6.24

Pebble mill to 7+ N.S. grind.

B	**Versamid 115**	46.59
	Methoxy Propanol	74.97
	Xylene	155.04

Physical Properties:

P/B Ratio — 1.0/1.0
Solids (wt.) — 41.3%
Solids (vol.) — 30.8%
Equivalent Ratio — 1.0/1.0
Visc. (#2 Zahn Cup) — 30–40 s

High-Solids Air-Dry Enamel

(White)

Rutile Titanium Dioxide R-902	287.91
Kelpol 3710-X-80	115.16
Nuosperse 657	6.86

Exkin #2	0.71
Xylene	57.58

Disperse to 7+ N.S.

Kelpol 3710-X-80	334.69
Cobalt (6%)	2.99
Calcium (6%)	5.98
Zirconium (6%)	5.98
Xylene	187.00

Physical Properties:

P/B Ratio — 0.8/1.0
Solids (wt.) — 66%
Solids (vol.) — 52%
Visc. (#4 Ford Cup, @ 25 C) — approx. 50–55 s

Semigloss Enamel

Formula No. 1

(Ultra-Deep Base)

Pigment Grind:

Propylene Glycol	56.2
Water	66.6
Cellosize® QP-15,000	1.0
AMP-95	2.0
Nalco 2324	3.3
Nalco 2321	1.9
Cosan 101	2.1
Zopaque® RCL-9	40.0
Kaomer 350	25.0
Omyacarb® UF	50.0

Letdown:

Water	78.6
Ucar® Acrylic 518	583.1
Nalco 2321	1.9
Texanol®	23.8

Triton® GR-7M	1.0
Gafac LO-529	2.0
Ammonium Hydroxide (28% aq. sol'n.)	1.8
Acrysol® TT-935	8.8
Water	8.3

Physical Properties:

PVC — 16.1%
Solids (wt.) — 44.8%
Solids (vol.) — 34.4%
Visc. (equilibrated) — 87 K.U. (untinted), 92 K.U. (tinted)
pH (initial) — 8.2
Wt./Gal — 9.83 lb

No. 2

(Deep-Base)

Pigment Grind:	
Propylene Glycol	56.2
Water	66.6
Cellosize® QP-15,000	1.0
AMP-95	2.0
Nalco 2324	3.3
Nalco 2321	1.9
Cosan 101	2.1
Zopaque® RCL-9	75.0
Kaomer 350	25.0
Omyacarb® UF	50.0
Letdown:	
Water	85.3
Ucar® Acrylic 518	566.5
Nalco 2321	1.9
Texanol®	23.8
Triton® GR-7M	1.0
Gafac LO-529	2.0
Ammonium Hydroxide (28% aq. sol'n.)	1.8
Acrysol® TT-935	8.8
Water	8.3

No. 3

(Tint-Base)

Pigment Grind:

Propylene Glycol	56.2
Water	66.6
Cellosize® QP-15,000	1.0
AMP-95	2.0
Nalco 2324	6.6
Nalco 2321	1.9
Cosan 101	2.1
Zopaque® RCL-9	200.0
Kaomer 350	25.0

Letdown:

Water	96.6
Ucar® Acrylic 518	543.1
Nalco 2321	1.9
Texanol®	23.8
Triton® GR-7M	1.0
Gafac LO-529	2.0
Ammonium Hydroxide (28% aq. sol'n.)	1.8
Acrysol® TT-935	6.6
Water	6.2

Physical Properties:

PVC — 24.0%
Total Solids (wt.) — 45.8%
Total Solids (vol.) — 31.1%
Visc. (equilibrated) — 100–105 K.U.
pH (initial) — 8.2
Wt./Gal — 10.48 lb
Freeze/Thaw (3 cycles) — Pass

No. 4

(White)

Pigment Grind:

Propylene Glycol	95.0
Tamol® SG-1	12.0
Colloid 643	1.8
Zopaque® RCL-9	250.0

Letdown:

Water	116.5
Ucar® Acrylic 518	514.8
Dimethylethanolamine (50% aq.)	2.0
Colloid 643	1.8
Texanol®	19.8
Super-Ad-It	1.0
Triton® GR-7M	0.5
Cellosize® QP-15,000 (2.5% aq. sol'n.)	50.0
Ammonium Hydroxide (28% aq. sol'n.)	1.0

Physical Properties:

PVC — 23.0%
Total Solids (wt.) — 49.8%
Total Solids (vol.) — 34.4%
Visc. (equilibrated) — 100–110 K.U.
pH (initial) — 8.1
Wt./Gal — 10.75 lb
Freeze/Thaw (3 cycles) — Pass

No. 5

(White)

A	**Versamid 115-173**	142.1
	R-901 Titanium Dioxide	226.8
	Cellosolve®	96.3
	Isopropanol	28.8
	Xylene	23.8

B	**Epon® 1001**	192.9
	Zeolex 80	142.7
	N-Butanol	35.0
	Methyl Isobutyl Ketone	18.2
	Cellosolve®	103.00
	N-Propanol	38.9

Disperse materials in Part A to N.S. of 6 and Part B to N.S. of 5.5. Thoroughly mix Parts A and B and allow at least 30 min induction time before application. For air-spray application, coating viscosity should be reduced with **Cellosolve®** solvent.

Physical Properties:

Wt./Gal — 10.55 lb
Solids (wt.) — 63.53%
Solids (vol.) — 47.38%
P/B Ratio — 1.24:1
Typical Visc. — 67 K.U.
Typical 60° Gloss — 50

Gloss White Enamel

Formula No. 1

Pigment Dispersion:

Titanium Dioxide	262.0
Sag Control Agent	10.0
Modaflow® (sol'n.)	4.5
Aroflint 606	100.0
Solvent	11.0

Roller mill or pebble mill.

Clear Activator:

Aroflint 252	433.0
Solvent	26.0

Enamel Formula:

Pigment Dispersion	593.5
Clear Activator	459.0

Physical Properties:

PVC — 14.5%
Total Solids — 70.2%
Typical Visc. — 60–65 K.U.

No. 2

(Exempt)

A	**G-Cure 867-PWF-60**	394.6
	Ti-Pure® R-960	218.6
	Anti Terra U	3.3
	CAB 551.01	3.3
	Xylene	33.8
	Toluene	12.8
	Propylene Glycol Monomethyl Ether Acetate	165.3
B	**Desmodur N-75**	100.7
	Propylene Glycol Monomethyl Ether Acetate	67.8

Physical Properties:

PVC — 15.6%
P/B Ratio — 0.7
Solids (wt.) — 53.8%
Solids (vol.) — 40.0%
Visc. (#4 Ford Cup) — 25 s
Rule 66 status — Exempt

No. 3

(High-Performance, Architectural)

A	**Versaid 115-I73**	147.5
	Titanium Dioxide R-901	253.3
	Xylene	24.7
	Cellosolve®	99.9
	Isopropanol	29.9

B	**Epoxy Resin EEW-525**	200.2
	Zeolex 80	82.0
	n-Butanol	36.3
	Methyl Isobutyl Ketone	18.9
	Cellosolve®	106.9
	n-Propanol	40.3

Disperse materials in Part A to N.S. of 6 and Part B to N.S. of 5.5. Thoroughly mix Parts A and B and allow at least 30 min induction time before application. For air-spray application, coating viscosity should be reduced with **Cellosolve®** solvent.

Exterior Maintenance Enamel

(Dark Green)

X-3163 Iron Blue	16.8
X-1285 Iroquois Yellow	45.0
X-2846 Empress Yellow	6.9
GE SR-187 (30%)	526.0
Modaflow®	3.1
Nuosperse 657	2.5
Nuogel AO	3.0
Exkin #2	4.6
Cobalt Octoate (6%)	2.6
Manganese Octoate (6%)	1.6
Calcium Octoate (5%)	3.1
Mineral Spirits	170.0

Eggshell White Enamel

(For Walls and Woodwork)

Pigment Grind:

Water	128.6
Nuosept 95	2.4
Propylene Glycol	25.9
Tamol® 731	11.5
Patcote 801	1.9

Triton® N-57	1.0
AMP-95	1.0
Ti-Pure® R-900	250.0
Atomite®	150.0
Attagel® 50	2.5

Letdown:

Ucar® Acrylic 518	409.5
Texanol®	15.8
Propylene Glycol	38.3
Water	89.2
Ucar® Thickener SCT-200	22.5
Patcote 801	1.9
Triton® GR-7M	0.5
Ammonium Hydroxide (28% aq. sol'n.)	2.0

Physical Properties:

PVC — 47.0%
Total Solids (wt.) — 54.0%
Total Solids (vol.) — 35.8%
Visc. (equilibrated) — 90–95 K.U.
pH (initial) — 9.0 min.
Wt./Gal — 11.51 lb
Freeze/Thaw (3 cycles) — Pass

High-Solids Enamel

A	**Versamid 150**	176.43
	Cellosolve®	31.33
	n-Butanol	13.54
	Toluene	41.68
B	**Epoxy Resin EEW-190**	294.90
	C_{12}–C_{12} Aliphatic Glycidyl Ether	32.70
	Titanium Dioxide R902	100.00
	Barytes #1	220.00
	Asbestine 325	189.90
	Bentone® 27	42.60

Methanol/Water (95:5)	16.50
Silicone Resin SR82	9.00
Cellosolve®	59.75

Mix Ratio: Part A to Part B, 1:2 by vol.

Physical Properties:

Parts/100 wt. of Epoxy Resin — 54 phr
Total Solids (wt.) — 86%
Solids (vol.) — 78%

Clear Enamel

(Rule 66 Exempt)

A	**Epon® 1001**	238.0
	Silicone Resin Sol'n.	6.0
	Isopropanol	47.0
	Toluene	32.0
	2-Nitropropane	70.0
	Cellosolve®	40.0
B	**Versamid 115-I73**	178.0
	Special Naphtholite (66/3)	88.0
	Butanol	19.0
	Toluene	29.0
	Xylene	18.0
	Cellosolve®	18.0
	Isopropanol	9.0

Thoroughly mix Parts A and B prior to use. Allow 15–30 min induction time before application. Reduce viscosity as needed after induction period.

Physical Properties:

Wt./Gal. — 7.91 lb
Total Coating Solids (wt.) — 46.45%
Total Solids (vol.) — 40.39%
Typical Coating Visc. (#4 Ford Cup) — 34 s
Typical Tack-Free Time — 5 h
Typical Hard-Dry Time — 13 h

Exterior Trim Enamel

Formula No. 1

(Ultra-Deep Base)

Pigment Grind:	
Propylene Glycol	56.2
Water	66.6
Cellosize® QP-4400	2.0
AMP-95	2.0
Colloid 226-35	3.3
Colloid 643	1.8
Troysan 192	2.0
Ti-Pure® R-900	40.0
Atomite®	75.0
Letdown:	
Water	38.1
Ucar® Acrylic 518	583.1
Colloid 643	2.6
Texanol®	23.8
Polyphase AF-1*	7.2
Triton® GR-7M	1.0
Gafac LO-529	2.0
Cellosize® QP-4400 (2.5% aq. sol'n.)	50.0
Ammonium Hydroxide (28% aq. sol'n.)	1.8

Physical Properties:

PVC — 16.0%
Total Solids (wt.) — 44.7%
Total Solids (vol.) — 34.4%
Visc. (equilibrated) — 82 K.U. (untinted), 87 K.U. (tinted)
pH (initial) — 8.3
Wt./Gal — 9.84 lb

* A typical value. Adjust mildewcide for requirements of a specific region.

No. 2

(Tint-Base)

Pigment Grind:	
Methyl Carbitol®	56.2
Water	66.6
Cellosize® QP-4400	1.8
AMP-95	2.0
Colloid 226-35	6.6
Colloid 643	1.8
Troysan 192	2.0
Ti-Pure® R-900	200.0
Atomite®	25.0
Letdown:	
Water	52.5
Ucar® Acrylic 518	543.1
Colloid 643	2.6
Butyl Carbitol®	23.8
Polyphase AF-1*	7.2
Triton® GR-7M	1.0
Gafac LO-529	2.0
Cellosize® QP-4400 (2.5% aq. sol'n.)	50.0
Ammonium Hydroxide (28% aq. sol'n.)	1.8

Physical Properties:

PVC — 22.1%
Total Solids (wt.) — 48.5%
Total Solids (vol.) — 33.1%
Visc. (equilibrated) — 100–105 K.U.
pH (initial) — 8.3
Wt./Gal — 10.67 lb
Freeze/Thaw (3 cycles) — Pass

* A typical value. Adjust mildewcide for requirements of a specific region.

No. 3

(White)

Pigment Grind:

Water	146.0
Cellosize® QP-4400	3.5
Colloid 226-35	7.7
Igepal® CO-630	2.0
Troysan 192	2.0
Propylene Glycol	40.0
Colloid 681-F	0.9
Tronox CR-800	250.0

Letdown:

Ucar® Acrylic 518	506.6
Texanol®	25.3
Colloid 681-F	1.9
Triton® GR-7M	1.0
Propylene Glycol	70.0
Polyphase AF-1*	7.2
Ammonium Hydroxide (28% aq. sol'n.)	2.7

Physical Properties:

PVC — 13.2%
Total Solids (wt.) — 43.2%
Total Solids (vol.) — 34.3%
Visc. (equilibrated) — 82 K.U. (untinted), 95 K.U. (tinted)
pH (initial) — 8.3
Wt./Gal — 9.57 1

---

* A typical value. Adjust mildewcide for requirements of a specific region.

Exterior Flat Paint

Formula No. 1

(White Acrylic)

Propylene Glycol	20.0
Natrosol® 250 MHR (2%)	100.0
Potassium Tripolyphosphate	1.5
Arco™ Dispersant 201 (33.7%)	10.6
Colloid 643	1.0
Methyl Carbitol®	35.0
Water	72.0
Ti-Pure® R-902	225.0
Minex 4	140.0
Icecap K	50.0
AZO-11	25.0
Attagel® 50	5.0

Disperse at high speed. Add:

Water	85.0
QR-708 (35.0%)	10.0
Butyl Carbitol®	16.0
Texanol®	10.0
Rhoplex® AC-64 (60.0%)	306.0
Colloid 643	3.0
Skane® M-8	2.0
Ammonium Hydroxide (28%)	2.0
Water	45.0

No. 2

Natrosol® 250 MHR (2.5%)	120.0
Ethylene Glycol	25.0
Propylene Glycol	34.0
Tamol® 850 (30%)	10.5
Potassium Tripolyphosphate	1.5
Triton® CF-10	2.5
Drew® L-475	1.0
Ti-Pure® R-902	262.5
Minex 4	190.2

Letdown:

Rhoplex® AC-64 (60.5%)	380.0
Drew® L-475	3.0
Texanol®	11.0
Super-Ad-It	9.0
Ammonium Hydroxide (28%)	2.0
Water and/or 2.5% **Natrosol® 250 MHR**	125.9

Physical Properties:

PVC — 40.0%
Total Solids (vol.) — 41.0%
Visc. (initial) — 80–85 K.U.
Visc. (equilibrated, sheared) — 81–91 K.U.
pH — 8.9–9.3

Exterior Sheen Paint

(White)

Water	65.0
Tamol® 850	14.2
Propylene Glycol	60.0
Colloid 643	2.0
Ti-Pure® R-902	250.0
ASP®-170	106.0

Letdown:

Rhoplex® AC-64 (60.5%)	351.7
Texanol®	10.1
Super-Ad-It	9.0
Colloid 643	4.0
Propylene Glycol	40.0
Water	29.8
Natrosol® 250 MR (2.5%)	167.5

Physical Properties:

PVC — 35.0%
Total Solids (vol.) — 35.0%
Visc. (initial) — 78–82 KU
Visc. (equilibrated) — 89–93 K.U. (hand stir), 78–82 K.U. (sheared)
pH — 9.0–9.5

House Paint and Tint Base

(White, Acrylic Latex)

Water	125.0
Ethylene Glycol	23.2
Cosan PMA-100	1.5
Potassium Tripolyphosphate	2.0
Tamol® 850	6.0
Triton® N-101	2.0
Defoamer	1.0
Zinc Oxide XX 503	50.0
Titanium Dioxide OR-600	250.0
Nyad® 400	50.0
Nycarb 20	100.0

High-speed dispersion for 15 min. Add:

Water	21.0
Attagel® 50	5.0

High-speed dispersion for 5 min. Add at slow speed:

Water	197.84
Natrosol® 250 HBR	4.00
Texanol®	11.88
Rhoplex® AC-64 (60.5% NV)	336.00
Defoamer	1.00

Physical Properties:

PVC — 40.0%
Total Solids (wt.) — 54.9%
Total Solids (vol.) — 37.0%
Visc. — 90 K.U.
pH — 9.2
Wt./Gal — 11.85 lb
Sheen 85° — 3

Weather Barrier Coating

Pigment Grind:

Water	143.3
Cellosize® ER-52M	4.0
Colloid 226/35	8.0
Potassium Tripolyphosphate	2.0
Ethylene Glycol	20.0
Surfynol® 104	1.2
Colloid 643	1.5
Ti-Pure® R-960	103.0
Atomite®	385.0
AZO-55	40.6
Nopcocide® N-96	5.0

Letdown:

Ucar® Latex 123	496.6
Texanol®	6.0
Colloid 643	2.0
Ammonium Hydroxide (28% aq. sol'n.)	2.5

Physical Properties:

PVC — 40.0%
Total Solids (wt.) — 68.5%
Total Solids (vol.) — 53.8%
Visc. (Stormer) — 110 K.U.

Exterior Topcoat

(Black)

A	**Versamid 280-B75**	280.0
	Genamid® 2000	20.0
	Thixatrol ST	10.0
	Hi-Fine 80	400.0
	R-960	5.0
	Butyl Alcohol	298.0
	Raven 14	18.0

B	**Epon® 815**	500.0
	Thixatrol ST	10.0
	Hi-Fine 80	250.0
	Super Hi-Flash Naphtha	251.0

Disperse pigments in both parts to give an N.S. fineness-of-grind reading of at least 5. Parts A and B are thoroughly mixed 1:1 by volume and allowed to induct approximately 30 min prior to use.

Physical Properties:

PVC — 26.4%
Total Solids (vol.) — 54.5%
Visc. — 65–76 K.U.
Wt./Gal — 10.1 lb
Drying Time (firm) — 6 h
Drying Time (hard) — Overnight
Pot Life — 6 h
60° Gloss — 10

Vinyl-Acrylic Emulsion Interior Flat Paint

Formula No. 1

Water		250.0
Natrosol® 250 MR	Premix	4.0
Ethylene Glycol	Premix	15.0
Tamol® 731 (25%)		10.0
Triton® N-101		2.0
Colloid 643		2.0
Dowicil® 75		2.0
Ti-Pure® R-931		250.0
Snowflake		175.0
Icecap K		75.0

Letdown:

Water	50.0
Vinyl-Acrylic Emulsion (55%)	260.0
Texanol®	12.6
Colloid 643	2.0
Natrosol® 250 MR (2.5%)	86.8

Physical Properties:

PVC — 56.5%
Total Solids (vol.) — 33.9%
Visc. (initial) — 77–82 K.U.
Visc. (equilibrated, sheared) — 80–87 K.U.
pH — 8.0–8.5

No. 2

(Premium Quality)

Water	70.0
Natrosol® 250 MHR (2.5%)	75.0
Tamol® 731	11.6
Triton® CF-10	2.0
Nopco® NDW	2.0
Ethylene Glycol	25.0
Super-Ad-It	1.0
Ti-Pure® R-902	293.0
Icecap K	100.0
Silica 1160	39.3
Propylene Glycol	35.0

Letdown:

Vinyl Acrylic Emulsion (55%)	433.6
Texanol®	16.7
Natrosol® 250 MHR (2.5%)	74.9

Physical Properties:

PVC — 38%
Total Solids (vol.) — 40%
Visc. (equilibrated) — 81 K.U.
pH — 7.3

Interior Eggshell Paint

Propylene Glycol		50.0
Natrosol® 250 GR	} Premix	2.0
Tamol® 731		10.0

Water	50.0
Triton® CF-10	2.0
PAG-188	2.0
Dowicil® 75	1.0
Ti-Pure® R-900	250.0
Imsil® A-15	75.0

Letdown:

Rhoplex® AC-417 (48%)		422.0
Texanol®		15.0
Propylene Glycol		13.0
PAG-188		4.0
Triton® N-57		4.0
Acrysol® RM-4 (30%)	Premix	14.0
Water	Premix	26.3
Ammonium Hydroxide	Premix	2.5
Water		88.8
Natrosol® 250 GR (4% sol'n.)		65.5

Physical Properties:

PVC — 34.0%
Total Solids (vol.) — 31.6%
Visc. (initial) — 80–85 K.U.
pH — 8.8–9.3
60° Gloss — 15–20
85° Sheen — 30–40

Interior Semigloss

(White, Acrylic)

Propylene Glycol	75.0
Arco™ 201 (33.7%)	7.6
Deefo 495	2.5
Ti-Pure® R-900	255.0

Disperse at high speed. Add:

Rhoplex® AC-417 (48.0%)	510.0
Acrysol® RM-3 (30.0%)	34.0

DMAE 1:1 in Water		2.0
Deefo 495		2.5
Propylene Glycol		20.0
Butyl Carbitol®	Premix	23.0
Texanol®	Premix	7.7
Super Ad-It	Premix	1.0
Triton® GR-7M	Premix	0.5
Water		62.0
Natrosol® 250 MR (2.5%)		62.0

One-Coat Flat Wall Paint

Pigment Grind:

Water	290.4
Cellosize® QP-4400	5.0
Cosan 101	2.1
Tamol® 731	4.6
Potassium Tripolyphosphate	1.0
AMP-95	1.0
Triton® N-101	2.0
Ethylene Glycol	28.0
Patcote 803	1.9
Ti-Pure® R-931	250.0
Optiwhite	100.0
Imsil® A-15	75.0

Letdown:

Ucar® Acrylic 518	372.0
Texanol®	15.8
Patcote 803	1.9

Latex Stipple and Texture Interior Coating

(Polyethylene)

Water	33.0
Daxad® 30 (25%)	1.0
Everflex GT® (55%)	14.0
Short Stuff® Polyethylene 13 038F	2.0

Mica (325 mesh)	4.0
Methocel® 65HG	0.5
Duramite®	33.0
Ti-Pure® R-900-01	4.0
Celite® 281	8.0
PMA®-30	0.5

Add one-half of the formula water to mixer. Begin agitation; add **Daxad® 30, Everflex GT®, Short Stuff® 13 038F** and mica slowly and mix. Dry mix **Methocel® 65HG** with a portion of **Duramite®** and add this with the remainder of **Duramite®** to the mix. Continue agitation until smooth. Add **Ti-Pure® R-900-01** and mix until smooth. Slowly add the remainder of formula water followed by **Celite® 281** and **PMA®-30**. Continue mixing until a homogeneous mix is obtained.

Application: Brush, sponge or a wad of paper

Note: This coating when applied to drywall, plaster, cement, or block wall substrates will provide a very attractive stipple or textured appearance. It should be applied by bursh in heavy coats at full weight. The coating, while still wet, should be hand finished with a wad of paper, brush, or damp sponge in a uniform manner to obtain the desired stipple or texture effect.

A stippled or textured coating contains silicates such as micas that are suspended in the coating medium. When this type of coating is applied, the silicates impart a textured or stipped appearance to the surface being covered.

Physical Properties:

Appearance — Excellent, extra white
PVC — 73.5%
Total Solids (wt.) — 59.2%
Total Solids (vol.) — 38.8%
Visc. (@ 25 C) — 58,000 cps
Wt./Gal — 12.52 lb
Drying Time (set-to-touch) — 1 h brush on, 2 h textured

Epoxy Coating

A	Epoxy Resin*	100
	Titanium Dioxide	4
	Colored Pigments	2
	Blanc Fixe	34
	Quartz Sand (0.1–0.3 mm)	66
B	IPD-TMD-blend SE	58

Medium Gloss Topcoat Lacquer

(For Forced Air Dry)

Rhoplex® WL-93 (34.0%)		552.8
Water**		68.8
Butyl Cellosolve	Premix	112.8
Santicizer 160	Premix	14.1
BYK® 301	Premix	0.5
Dow Corning® P.A. #14	Premix	0.5
Water		90.6
FP-2 (15% **Syloid 166**)		26.9

Add in order listed, under agitation. Adjust viscosity with 14% ammonia to 200 ± 20 cps.

Physical Properties:

Total Solids (wt.) — 24 ± 1%
Total Solids (vol.) — 22 ± 1%
Visc. — 200 ± 20 cps
pH — 8.2–8.6
Wt./Gal — 8.42 ± 0.1 lb

* Bisphenol A-type with epoxy equivalent weight of about 190

** In geographical locations noted for troublesome water conditions such as unusual ionic concentration, deionized water may be desirable.

Nitrocellulose-Urethane Lacquer

Vehicle Solids (100% solids basis):	
UG RS Nitrocellulose ($^1/_4$ s)	43.0
Multron R-221	26.0
Desmodur® N	31.0
Solvent (urethane grade):	
Toluene	15.0
Exxate 600	10.0
Methyl Ethyl Ketone	40.0
Methyl Isobutyl Ketone	35.0
Catalyst:	
Zinc Octoate Sol'n.*	0.1

Nitrocellulose Lacquer

RS Nitrocellulose (30–35 cP)	50.0
Vynathene EY 902-30	50.0
Isopropyl Alcohol	21.5
n-Butyl Alcohol	158.5
n-Butyl Acetate	307.2
n-Propyl Acetate	35.7
Isopropyl Acetate	17.1
Toluene	360.0

Physical Properties:

Nonvolatiles — 10%

Visc. (#4 Ford Cup, @ 75 F) — 17 s

* Containing 8% zinc per 100 parts of solids (100% solids basis).

Food Container Coating

(Epoxy)

A	**Liquid Epoxy Resin EEW-190**	52.0
	Inorganic Pigment Colorant	15.0
	Barytes (bleached)	11.0
	Emtal® 41	10.0
	Icecap K	5.0
	Bentone® 27	2.0
	Ethanol	5.0
B	**Versamine 170**	11.0

Physical Properties:

Nonvolatiles — 95.5%
P/B Ratio — 0.68/1
Pot Life (@ 20 C) — 20 min
Through-Cure Time (@ 20 C) — 15 h

Paper, Film, Foil, and Inks Coatings

Formula	No. 1	No. 2	No. 3	No. 4	No. 5	No. 6	No. 7
1,4-Cyclohexanedimethanol-polyol	28.0	23.0	30.0	—	—	—	—
Niax LHT-240	7.0	12.0	12.0	—	—	—	—
Aroplaz 2575 (60% solids)	—	—	—	40.0	—	—	—
Hycar® 2600X138 (48% solids)	—	—	—	—	82.5	—	—
Celanese PD-4070 (69% solids)	—	—	—	—	—	—	37.2
CAP-504-0.2	—	—	—	—	—	7.8	—
Diisopropanolamine (50%)	—	—	—	—	to pH 9.0	—	—
Resimene U-980 (100% solids)	65.0	65.0	—	—	—	15.7	—
Resimene U-975 (100%)	—	—	—	16.0	2.0	—	—
Resimene 717 (84%)	—	—	—	—	—	—	9.0
Resimene 747 (100%)	—	—	58.0	—	6.0	—	—
BYK®-301 (50%)	—	—	—	—	—	1.0	—
Xylene/Ethanol (60:40)	—	—	—	37.2	—	—	—
Petrolatum (10% in xylene)	—	—	—	2.8	—	—	—
Isopropanol	—	—	—	—	3.0	—	—
Pluronic L62	—	—	—	—	0.5	—	—
n-Propyl Alcohol	—	—	—	—	—	53.3	—
Water	—	—	—	—	4.4	22.2	—
Deionized Water	—	—	—	—	—	—	53.8
Add at time of use:							
K-Cure 1040 (40% solids)	variable	variable	variable	4.0	variable	3.0	—

Furniture Finish Wiping Stains

Add, in order, under agitation:

		Formula No. 1 (Dark Walnut)	No. 2 (Golden Oak)
Rhoplex® WL-93 (34%)		342.6	342.6
Butyl Cellosolve		70.0	70.0
Santicizer 160		8.8	8.8
Deefo 806-102		1.0	1.0
Water*		69.5	69.5
Water	Premix	330.6	330.6
Calco Nigrosine 02P PDR	Premix	21.8	3.4
Calco CID Orange Y Ex. Conc.	Premix	7.5	6.9
Calco CID Crocein Scarlet MOS Conc.	Premix	3.4	—
Calco CID Tartrazine DB1 Conc.	Premix	3.4	24.1

Physical Properties:

	Formula No. 1	No. 2
Density (lb/gal)	8.58	8.57
pH	7.7	7.3
Visc. (cps)	20	20
Solids (% by wt.)	18.8 ± 1	18.6 ± 1
Solids (% by vol.)	16.2 ± 1	16.2 ± 1

Note: Wipe directly onto wood, or spray (as is) onto wood, immediately wipe off excess.

* In geographical locations noted for troublesome water conditions such as unusual ionic concentrations, deionized water may be desirable.

Semitransparent Stain

(Oil-Based)

Boiled Linseed Oil	9.4
Bentone® 38 Gel (13.5%)	1.9
Red Iron Oxide VVF-B293-F	1.9

Disperse on Cowles Dissolver. Then add at lower speed:

Aromatic 150	35.3
Paraffin Wax	1.1
Boiled Linseed Oil	47.3
Busan 1009	3.1

Note: Low-melting paraffin wax (melting points of 123–125 F) should be used. The paraffin wax should first be melted, then added to the **Aromatic 150** under strong agitation. Separation of the paraffin may occur at low temperatures. Heating to room temperature under agitation should resolubilize the material.

UV-Curable Wood Coating

Hexanediol Diacrylate	37
Photomer® 3016	30
Ethoxyethoxyethyl Acrylate	15
Stalink® 106	10
N-Vinyl-2-Pyrrolidone	5
Irgacure® 184	3

Physical Properties:

Visc. — 2300 cps
Cure Conditions — 30 fpm per 200 w/in.

Clear Finish

Formula No. 1

(Long-Oil Alkyd)

Cargill BB-670	13.69
Pine Oil	1.03
Manganese Naphthenate (6%)	0.50
Mineral Spirits	28.63
Solvesso 150	54.15
Busan 1009	2.00

No. 2

(Industrial Vinyl Toluene-Alkyd)

Xylol	75.6
Paraffin Wax	2.0
Chempol 13-2403 (60% solids)	16.9
Manganese Naphthenate (6%)	0.1
Zirco Drier (6%)	0.3
Busan 1009	5.1

Note: The paraffin wax should first be melted prior to addition to the xylol under strong agitation. Separation of the paraffin may occur at low temperatures. Use of aliphatic cosolvents such as hexane may improve paraffin stability.

Polyurethane Coatings

		Formula No. 1 *(White Pigmented)*	No. 2 *(Clear)*
A	Millbase, high speed disperser:		
	Ektasolve EE Acetate } Resin cut	63.0	150.0
	Toluene } Resin cut	27.0	65.0
	CDX-587	135.0	322.0
	n-Butyl Acetate	4.0	104.0
	Zopaque® RCI-6	247.0	—

Letdown:			
	Ektasolve EE Acetate } Resin cut	68.0	—
	Toluene }	29.0	—
	CDX-587	146.0	—
	n-Butyl Acetate	135.0	—
	Dow Corning® Paint Additive #57	0.4	0.4
B	**Desmodur® N-75**	119.0	137.0
	n-Butyl Acetate	55.0	73.0

Mix Parts A and B under mild agitation and dilute to spray viscosity with *n*-butyl acetate.

	No. 3	No. 4
PCP-0300	144.0	144.0
Quadrol	14.4	14.4
Titanium Dioxide	100.0	100.0
Green Oxide	25.0	25.0
Silicone 200	2.0	2.0
E-448	169.0	169.0
Viplex 525	150.0	—
Dibutyl Phthalate	—	150.0
Physical Properties:		
Ultimate Load (lb)	37.90	11.5
Tensile Strength (psi)	1170	443
Elongation @ break (%)	95	55

Note: Formula No. 3 with **Viplex 525** shows three times better tensile and about 2 times better elongation than Formula No. 2 with DBP.

Spar Varnish

Formula No. 1

Ucar® Phenolic Resin CK-2500	100.00
Butanol	15.00
Mineral Spirits	85.00
Tung Oil	200.00

Dissolve **CK-2500** in butanol and mineral spirits. Add tung oil slowly while stirring.

Physical Properties:
Visc. — 180 cP
Nonvolatiles (wt.) — 75%
Nonvolatiles (vol.) — 70%

No. 2

(Cold-Blend, for High-Solids and Wrinkle Resistance)

Ucar® Phenolic Resin CK-2400	100
Tung Oil	174
Linseed Oil (alkali-refined)	50
Mineral Spirits	146
Isopropanol	15

Prepare a 40% NV solution by dissolving 100 parts **CK-2400** in 135 parts mineral spirits and 15 parts isopropanol. Add the raw tung oil resin solution slowly with stirring. When the resin is dispersed in an equal weight of oil, add the remaining ingredients rapidly.

Physical Properties:
Tung/Linseed — 75/25%
Sp. Gr. — 0.916
Wt./Gal — 7.62 lb
Nonvolatiles (wt.) — 65%
Visc. (Gardner-Holdt) — D
Color (Gardner 1933) — 8–9

Water-Repellent Wood Preservative

Paraffin Wax	0.8
Boiled Linseed Oil	8.7
Solvesso 150	88.5
Busan 1009	2.0

Melt the paraffin wax. Then slowly add under vigorous agitation to the **Solvesso 150**, which should be at room temperature. Add the boiled linseed oil and **Busan 1009**. Mix until homogeneous. Separation may occur at low temperature. Warming to room temperature under agitation will redissolve the ingredients.

Latex Traffic Line Coating

(Polyethylene)

A	**Ti-Pure® R-900-01**	14.5
	Mississippi® M-60	14.5
	Nytal® 300	6.0
	Celite® 281	5.0
	Short Stuff® Polyethylene 13 038F	2.0
	Mica (325 mesh, WG)	6.0
B	**Kelecin® F**	0.7
	Keltrol® 1074	32.0
C	**Intercar® Cobalt** (6%)	0.3
D	Toluol	19.0

Blend **Ti-Pure® R-900-01** and **Mississippi® M-60**. Add **Nytal® 300** and mix. Add **Celite® 281** and mix for 5 min. Add one-half of the **Short Stuff® Polyethylene 13 038F** and mix. Add the mica and mix. Mix all dry ingredients thoroughly. Preblend Part B; mix until smooth. Blend Part A and mix with Part B to a smooth consistency. Add the remainder of **Short Stuff® Polyethylene 13 038F** and mix for 5 min, slowly adding Part D; mix until homogeneous. Add Part C and mix for 10 min.

Physical Properties:

PVC — 46.2%
Vehicle Solid — 37.2%
Total Solids (wt.) — 67.5%
Visc. (Spindle #5, 20 rpm, 24 C) — 24,000 cps
Wt./Gal — 11.5 lb

Phenolic/Epoxy Powder Coatings

	Formula No. 1	No. 2	No. 3
Epon® Epoxy Resin 1004	25.0	60.0	—
Epon® Epoxy Resin 1007	—	—	75.0
Ucar® Phenolic Resin BK-5918	75.0	40.0	25.0
Modaflow® (powder)	1.0	1.0	1.0
Tullanox 500	0.5	0.5	0.5

Dry-blend ingredients on a Hobart mixer. Extrusion-mix dry blend at 220 F (104 C). Cool with liquid nitrogen and grind at 0 F (−18 C) through a 0.010-in. screen in a hammer mill. Vacuum dry overnight at room temperature and then screen through a 140 or 325 mesh* screen.

Note: Electrostatic spray. Bake 20 min at 350 F (177 C).

Asphalt Cutback Roof Coatings

Formula No. 1

(Black)

Asphalt Solvent Cutback (65% solids)	87.0
Bentone® SD-1	0.5
Talc (5-micron average size)	5.5
Talc (16-micron average size)	5.5
Pulpex® E-DH	1.5

Add **Bentone® SD-1** slowly to solvent cutback; stir for 15 min. Add **Pulpex® E-DH**; mix until strong uniform gel is obtained. Add talc; mix thoroughly until smooth.

Physical Properties:

Visc. (Brookfield, #7 spindle, 20 rpm, 73 F) — 53,000 cps

* best appearance

No. 2

(Fibrated Aluminized)

Asphalt Solvent Cutback (65% solids)	64.0
Bentone® SD-1	0.5
Talc (5-micron average size)	5.5
Talc (16-micron average size)	5.5
Pulpex® E-DH	1.5
Aluminum Paste (coarse grade)	22.0
Syloid® AL-1	1.0

Add **Bentone® SD-1** slowly to asphalt cutback and stir for 10 min. Add **Syloid® AL-1** stabilizer and mix until well dispersed, about 2 min. Add **Pulpex® E-DH** and mix for 20 min; then add two grades of talc and mix for 5 min. Aluminum paste is added last and mixed in only another 5 min. Do not overmix paste.

Physical Properties:

Reflectance (3 days, 96 F, 100% r.h.) — 54%
Visc. (Brookfield, #7 spindle, 20 rpm, 73 F) — 54,000 cps

Asphalt Emulsion Roof Coating

Water	31.0
Attagel® 40	3.0
Pulpex® E-DH*	1.0

Mix well for 15 min or until smooth and fibers are totally dispersed. Stop agitation, then add:

Asphalt Emulsion (60% total solids; anionic SS1H type)	32.0

* The dry-fluff form is cited here; however, the wet-fluff form is sufficient to produce uniform dispersion at lower cost. Use 2% **Pulpex® E-DV** wet-fluff, as is, since **E-DV** is nominally 50% total solids.

Start agitation; mix well and add:

Carbon Black Sol'n.	0.4
Emulsion Stabilizer	optional
Inert Fillers ($CaCO_3$, talc, slate, bentonite, etc.)	32.6

In handling asphalt emulsions, care should be exercised in the order of addition and temperature at which finished product is applied—not to exceed 100 F. Up to 25% additional water may be used for dilution when applied. Prior dilution may cause settling of fillers.

Mobile Home Roof Coating

	Formula No. 1 *(Nonfibered)*	No. 2 *(Fibered)*
Pigment Grind:		
Water	368.3	362.0
Cellosize® QP-15,000	6.0	6.0
Phenyl Mercuric Acetate (100%)	1.0	1.0
Tamol® 960	13.1	13.1
Potassium Tripolyphosphate	2.0	2.0
Colloids 681-F	0.9	0.9
Tergitol® NP-10	2.2	2.2
Ethylene Glycol	18.6	18.6
Zopaque® RCL-9	200.0	200.0
AZO-55LO	25.0	25.0
Mica (325 mesh W.G.)	25.0	—
Nyad® G	—	100.0
ASP®-170	50.0	—
Attagel® 50	5.0	5.0
Letdown:		
Ucar® Latex 376	362.0	362.0
Texanol®	9.9	9.9
Colloids 681-F	1.8	1.8

Asphalt Cutback Underbody Coating

Asphalt Cutback	60.0
Arquad® 2C-75	1.4

Mix 2 min at slow speed*

Attagel® 36	12.0

Mix 10 min at medium speed*

Asphalt Cutback	13.3
Nyad® G	13.3

Mix 15 min at medium speed.*
Note: Formulated to meet Fed. Spec. TT-C-520B

Physical Properties:
Clay/Surfactant Ratio — 8.5/1
Mineral Content (wt.) — 25.3%

Typical Asphalt Properties:
Ring & Ball Softening — 175
Needle Penetration — 25–26
Cutback Solids (wt.) — 50%
Saybolt Visc.(@ 77 F) — 100 s

Paint Stripper

Methylene Chloride	90.6
Methocel® HB	1.4
Methanol	3.0
Monamulse 947	5.0

* Mixer: electrically driven laboratory paddle mixer.

Disperse **Methocel® HB** in the methylene chloride with moderate agitation. Add methanol and increase agitation. After system turns translucent, add **Monamulse 947**, and stir until completely dissolved.

Heavy-Duty Tank Linings

A	**Epoxy Resin EEW 525** (80% solids in MEK)	98.9
	Epoxy Resin EEW 525 (55% solids in MIBK/toluene)	182.0
	Silicone Resin SR-82	3.3
	Nuosperse 657	4.5
	Methyl Ethyl Ketone	64.5
	Methyl Isobutyl Ketone	37.3
	Toluene	144.0
	R-960	37.8
	Barytes No. 1	182.0
	Emtal® 41	184.00
	Bentone® 18C/Denatured Ethyl Alcohol Slurry	8.9
B	**Versamine A51**	67.8
	Methyl Isobutyl Ketone	26.1
	Butyl Cellosolve®	2.9
	Toluene	29.0

Slowly add the curing agent to the base component with good agitation until thoroughly dispersed. Allow the mixture to induct 30–60 min before application.

Physical Properties:

Nonvolatiles (wt.) — 58.3%
P/B Ratio (wt.) — 1.9/1.0
PVC — 41.1%
Visc. — 68 K.U.
Wt./Gal — 10.7 lb

Chapter VI
DETERGENTS

Stain Remover

Water	58.55
Igepal® CO-630	25.00
Westvaco 1550	10.00
Diethanolamine	3.70
Monamid® 705	2.00
Monaquat P-TC	0.75

Pump-Spray Laundry Prespotter

Formula No. 1

Neodol® 25-3	10.0
Neodol® 23-6.5	10.0
Shell Sol® 71	20.0
Isopropyl Alcohol	12.0
Triethanolamine Oleate*	3.5
Water	q.s. 100.0

No. 2

Neodol® 25-3	22.0
Shell Sol® 71	18.0
Ultra® SXS (40% AM)	30.0
Triethanolamine Oleate*	2.5
Water	q.s. 100.0

* Can be prepared *in situ* from triethanolamine and oleic acid.

Pump-Spray Laundry Soil and Stain Remover

	Formula No. 1	No. 2	No. 3	No. 4
Neodol® 25-7	3.9	—	5.2	—
Neodol® 25-9	—	3.9	—	5.2
Neodol® 25-3	2.1	2.1	2.8	2.8
Sodium Xylene Sulfonate (40% AM)	10.0	7.5	10.0	7.5
Ethylenediamine Tetraacetic Acid, Tetrasodium Salt	0.5	0.5	0.5	0.5
Water, Dye, and Perfume	q.s.	q.s.	q.s.	q.s.

Physical Properties:

Visc. (cps)	7	7	9	9
Clear Pt. (°F)	62	63	66	66

	No. 5	No. 6	No. 7	No. 8
Neodol® 25-7	6.5	—	9.75	—
Neodol® 25-9	—	6.5	—	9.75
Neodol® 25-3	3.5	3.5	5.25	5.25
Sodium Xylene Sulfonate (40% AM)	12.5	7.5	15.00	7.50
Ethylenediamine Tetraacetic Acid, Tetrasodium Salt	0.5	0.5	0.5	0.5
Water, Dye, and Perfume	q.s.	q.s.	q.s.	q.s.

Physical Properties:

Visc. (cps)	10	17	15	18
Clear Pt. (°F)	66	67	66	69

For Formulas No. 1—8:

Load the two **Neodols®** into the vessel and mix. Add the sodium xylene sulfonate solution and mix. Add the water and EDTA salt, either premixed or in that order. Mix until homogeneous.

Notes: Heating the water and/or **Neodols®** to 100–120 F before addition will facilitate solution, but is not essential.

pH on all samples is approximately 10, due to presence of the highly alkaline EDTA salt used. If a lower pH, e.g., 7–7.5, is preferable, this can be obtained by using the appropriate EDTA salt and/or neutralizing with $NaHSO_4$.

Light-Duty Liquid Detergent

	Formula No. 1	No. 2	No. 3	No. 4	No. 5
Neodol® 25-3A (60% AM)	22.7	18.3	12.5	8.3	5.5
Sulframin® 1260	36.7	30.0	25.0	13.5	9.2
P&G Amide 72	5.0	4.0	2.5	1.9	1.2
Ultra SXS (40% AM)	5.0	8.5	—	3.0	3.0
Ethanol*	5.0	2.7	1.8	1.2	0.8
Sodium Sulfate**	2.6	2.2	0.8	0.6	0.4
Sodium Chloride	—	—	—	1.0	1.5
Urea	—	—	4.3	—	—
Water, Dye, and Perfume	q.s.	q.s.	q.s.	q.s.	q.s.
Physical Properties:					
Total Active Matter (%)	40	33	25	15	10
Visc. @ 25 C (cps)	200	200	205	220	125
Clear Pt. (°C)	+4	−5	0	−1	−1
Foam Performance†	128	109	81	62	47

Heavy-Duty Liquid Laundry Detergent

Formula No. 1

(Citrate-Built)

Neodol® 25-7	4.5
Neodol® 25-9	5.7
Witconate 1250	18.1
Sodium Xylene Sulfonate	3.8
Triethanolamine	4.5
Trisodium Citrate	9.1
Ethanol	4.5
Water	49.8

* Derived from Neodol® 25-3A except formulation when 1.8% added.

** None added when oleum-derived LAS is used.

† Measured relative to the Shell SM-1 standard as 100.

Note: Sodium citrate is used as a builder to soften water in laundry detergents. It sequesters the calcium and magnesium in water which normally will interfere with the detergent action of the surfactants. The citrate also acts as an anti-redeposition agent.

	No. 2	No. 3	No. 4
	(Unbuilt, 30% Active)		
Neodol® 91-6	22.5	—	11.25
Neodol® 91-8	—	22.5	—
Neodol® 25-7	—	—	11.25
Sulframin® 1260 (100% basis)	7.5	7.5	7.5
Carbonate Mix*	1.5	2.0	2.0
Tinopal® Mix**	0.3	0.3	0.3
Water, Dye, and Perfume	q.s.	q.s.	q.s.
Physical Properties:			
Visc., 23 C (cps)	100	120	150
Clear Pt. (°C)	–1	–1	1
Clear Pt. (°F)	31	31	33
Pour Pt. (°C)	–3	–4	–2
Pour Pt. (°F)	27	25	28
pH	9.8	9.8	9.8
Freeze-Thaw Test† (3 cycles)	++	+	+
High Temp. Stability† (120 F, 1 week)	++	++	++

	No. 5	No. 6	No. 7
	(Unbuilt, 30% Active)		
Neodol® 91-6	11.25	—	11.25
Neodol® 91-8	—	22.5	—
Neodol® 25-7	—	—	11.25

* Equal weights of sodium carbonate and sodium bicarbonate.

** **Tinopal® 5 BM** (90%) plus **Tinopal® RBS 200%** (10%)

† + is passes, each + represents a test on a separate preparation.

Neodol® 25-12	11.25	—	—
Sulframin® 1260 (100% basis)	7.50	7.5	7.50
Carbonate Mix*	2.50	1.5	2.00
Ethanol SD-3A	2.00	—	2.00
Tinopal® Mix**	0.30	0.3	0.30
Water, Dye, and Perfume	q.s.	q.s.	q.s.
Physical Properties:			
Visc., 23 C (cps)	125	120	125
Clear Pt. (°C)	−1	1	−1
Clear Pt. (°F)	30	33	30
Pour Pt. (°C)	−5	−3	−4
Pour Pt. (°F)	23	27	25
pH	9.7	9.8	9.8
Freeze-Thaw Test† (3 cycles)	++	+	+
High Temp. Stability† (120 F, 1 week)	++	++	++

Hand Wash Liquid Detergent

	Formula No. 1	No. 2
		(Cold Water)
Neodol® 25-7	—	10
Neodol® 25-3S (60% AM)	25	0
Nitrene 6180	5	5
Monateric CAB		15
Water, Brightness, Perfume, etc.	q.s. 100	q.s. 100

No. 3

Water	70.0
Monamine ALX-100S	20.0
Surfonic® N-95	10.0

---

* Equal weights of sodium carbonate and sodium bicarbonate.

** Tinopal® 5 BM (90%) plus Tinopal® RBS 200% (10%)

† + is passes, each + represents a test on a separate preparation.

Add ingredients in the order listed. The pH level may be lowered to 7.0–7.5 with 85% phosphoric acid or 25% sulfuric acid.

Automatic Dishwashing Detergent

	Formula No. 1	No. 2
	(Powder)	*(Liquid)*
Sodium Tripolyphosphate	30–50	20–40
Sodium Metasilicate	5–18	—
Sodium Carbonate	40–60	10–50
Surfactant	0.5–3.0	0.5–3.0
Rheology Modifiers and Enzyme Stabilizers	—	5–20
Sodium Sulfate, Water, etc.	Bal. to 100	—
Water	—	Bal. to 100
Maxamyl® P 5,000	1–3	1–3
Maxacal® P 400,000 or **Maxatase® P 440,000**	1–3	1–3
Maxamyl® WL 7,000	—	1–3
Maxacal® L 300,000 or **Maxatase® LS 400,000**	—	1–3

No. 3

(Powder)

Surfonic® LF-17	2
Sodium Metasilicate (anhyd.)	37
Sodium Tripolyphosphate	37
Sodium Carbonate	25

No. 4

(Phosphate-Free)

Miranol® JEM Conc. or **Miranol® J2M Conc.**	.7
Nitrilotriacetic Acid	41.0
Sodium Metasilicate Pentahydrate	55.3

Sodium Chloroisocyanurate	2.0
Polyethylene Glycol	1.0

No. 5

(Institutional Liquid)

Miranol® JEM Conc.	3.0
Tetrapotassium Pyrophosphate	10.0
Sodium Metasilicate Pentahydrate	10.0
Water	77.0

General and All-Purpose Cleaner

	Formula No. 1	No. 2
Neodol® 25-12	10.0	4.0
Tetrapotassium Pyrophosphate	2.0	4.0
Trisodium Phosphate Dodecahydrate	—	2.0
Sodium Metasilicate Pentahydrate	4.0	—
Butyl Dioxitol® Glycol Ether	—	5.0
Sodium Xylene Sulfonate (40%)	4.0	5.0
Ammonium Hydroxide (28%0	—	1.5
Hexylene Glycol	6.0	—
Water	q.s.	q.s.
Physical Properties:		
Phase Coalescence Temp. (°C)	63	73
Visc. @ 25 C (cps)	8	5

Formula No. 1: The nonionic surfactant and glycol are dissolved in water. The TKPP is then stirred in.

Formula No. 2: Dissolve the phosphate salts in the water. Add the SXS, then the surfactants, then the butyl **Dioxitol®**. Stir until dissolved.

Suggested Use Directions: For light-duty cleaning use $^1/_4$ cup per gallon water. For heavy-duty cleaning, use 1 cup per gallon water.

Spray and Wipe Hard Surface Cleaner

	Formula No. 1	No. 2	No. 3
Neodol® 91-6	1.0	—	—
Neodol® 91-8	—	1.0	—
Neodol® 25-12	—	—	1.0
Butyl Oxitol® Glycol Ether	5.0	5.0	5.0
Tetrapotassium Pyrophosphate	2.5	2.5	2.5
Water	q.s.	q.s.	q.s.
Physical Properties:			
Phase Coalescence Temp. (°C)	41	51	61
Visc. @ 25 C (cps)	3	4	4
pH	—	10	10

Product intended for light-duty tasks such as cleaning hard surfaces, window panes, and tiles.

Use Dilution: 1 part in 4 parts water.

Note: Addition of 2% sodium xylene sulfonate (40% sol'n.) to Formula No. 1 increased the phase coalescence temperature to 52 C; 4% raised it to 68 C. Analogous results can be obtained with Formula No. 2.

	No. 4	No. 5	No. 6
Neodol® 91-6	3.0	—	—
Neodol® 91-8	—	3.0	—
Neodol® 23-6.5	—	—	3.0
Butyl Oxitol® Glycol Ether	3.0	3.0	3.0
Sodium Tripolyphosphate	2.0	2.0	2.0
Sodium Metasilicate Pentahydrate	2.0	2.0	2.0
Ethylenediamine Tetraacetic Acid, Tetrasodium Salt	0.5	0.5	0.5
Sodium Xylene Sulfonate (40% AM)	4.0	4.0	4.0
Water	q.s.	q.s.	q.s.
Physical Properties:			
Phase Coalescence Temp. (°C)	61	71	60
Visc. @ 25 C (cps)	5	5	6
pH	13	—	—

Dissolve STPP and silicate in warm water. Add EDTA salt, nonionic surfactant, SXS and **Butyl Oxitol®** with agitation until homogeneous.

	No. 7	No. 8	No.9
Neodol® 91-6	1.0	—	—
Neodol® 91-8	—	1.0	—
Neodol® 23-6.5	—	—	1.0
Butyl Dioxitol® Glycol Ether	5.0	5.0	5.0
Tetrapotassium Pyrophosphate	2.5	2.5	2.5
Trisodium Phosphate Dodecahydrate	4.6	4.6	4.6
Sodium Xylene Sulfonate (40% AM)	3.0	3.0	3.0
Water	q.s.	q.s.	q.s.
Physical Properties:			
Phase Coalescence Temp. (°C)	65	63	62
Visc. @ 25 C (cps)	4	4	5
pH	12.8	—	—

Dissolve the TKPP in the water. Add the SXS solution, then the nonionic surfactants and **Butyl Dioxitol®**. Mix until clear.

Oven Cleaner

Miranol® C2M-SF Conc.	3.0
Acrysol® ICS-1	6.0
Potassium Hydroxide (45%)	22.0
Water	69.0

Mix the water and **Acrysol® ICS-1** together. Slowly add the **Miranol® C2M-SF Conc.** Under high agitation add the potassium hydroxide very slowly to insure uniformity.

Toilet Bowl Cleaner

(Liquid)

Deionized Water	87.5
Avanel S-90	5.0
Hydrochloric Acid (37%)	7.5

Charge vessel with ingredients in the order listed using moderate agitation.

Acid Glass Cleaner

Formula No. 1

Ethylene Glycol Monobutyl Ether	10.0
Avanel S-70	0.3
Deionized Water	q.s. to 100.0
Color	as desired
Acetic Acid	to pH 3.5–4.0

Charge the ethylene glycol monobutyl ether, **Avanel S-70**, and most of the water. Dissolve the dye in the portion of the water charge held out and add to the product. Adjust pH to 3.5–4.0 with acetic acid.

Notes: On prolonged storage glycol ethers have been shown to form peroxy compounds which can adversely affect the dyes normally used in glass cleaners.

This formulation is also effective on surfaces other than glass normally found in the home such as counter tops and painted surfaces.

No. 2

Citric Acid (anhyd.)	0.500
Tergitol® NPX	1.500
Isopropanol	97.500
Sandocryl Blue B-RLE	0.003
Water	800.000–1000.000

Note: Citric acid is used along with a detergent and a solvent as a window glass cleaner. It is exceptionally good for removal of spots caused by water hardness due to the ability of citric acid to chelate calcium and magnesium. Oily smudges are also easily dissolved by the detergent and solvent.

Pet Stain Removal Carpet Cleaner

Citric Acid	10
Ultrawet 42K	5
Water	85

Directions: Saturate the stain area and brush to lather. Let stand 15 min. Brush to lather again and rinse with water and wiping with sponge.

Note: Citric acid is used in a formulation to treat pet urine stains on household carpets. Dog urine is generally about pH 8–9 which releases ammonia and other odors. Treating the stain with citric acid at pH 2 effectively eliminates the undesirable odors by forming their citrate salts which are not volatile. The residue can then be rinsed away with water.

Low-Foam Floor Cleaner

	Formula No. 1	No. 2	No. 3
	(For Machine Application)		
Neodol® 91-6	3.0	—	—
Neodol® 91-8	—	3.0	—
Neodol® 25-9	—	—	3.0
Triton® H-66	4.0	4.0	4.0
Tetrapotassium Pyrophosphate	8.0	8.0	8.0
Potassium Hydroxide (45%)	1.5	1.5	1.5
Water, Dye, and Perfume	q.s.	q.s.	q.s.
Physical Properties:			
Phase Coalescence Temperature (°C)	58	71	70
Visc. @ 25 C (cps)	5	4	6
pH	13.4	—	—

Application: Use concentration: 1–2 oz in 4 gal water.

Caution: This solution may discolor white vinyl tiles if allowed to remain for lengthy periods.

Wax Stripper

Formula No. 1

(Low Foaming)

Miranol® JEM Conc.	4.0
Tetrapotassium Pyrophosphate	4.8
Trisodium Phosphate	3.0
Starso	5.0
Dowanol EB	1.0
Water	82.2

No. 2

Miranol® J2M-SF Conc.	4.0
Monoethanolamine	3.0
Dowanol EB	5.0
Potassium Hydroxide (45%)	3.0
Tetrapotassium Pyrophosphate	3.0
Water	82.0

No. 3

Miranol® J2M-SF Conc.	4.2
Dowanol EB	3.5
Isopropyl Alcohol	2.5
Potassium Hydroxide (45%)	4.0
Monoethanolamine	5.0
Tetrapotassium Pyrophosphate	2.0
Trisodium Phosphate	2.0
Water	76.8

Chapter VII
DRUGS

Diethyl Carbamazine Citrate Tablet

(200 mg)

	mg/tablet
Diethyl Carbamazine Citrate	200.00
Emcompress®	150.00
Avicel® PH 101	72.00
Explotab®	22.50
Cab-O-Sil®	1.80
Magnesium Stearate	4.50

1 Mix diethyl carbamazine citrate, **Emcompress®**, and **Avicel®** together for 10 min.
2 Add **Explotab®** to (1) and mix thoroughly.
3 To (2) add **Cab-O-Sil®** and blend 10–15 min.
4 Add magnesium stearate to (3), mix for 5 min and compress.

Prednisone Tablet

(5 mg)

	mg/tablet
Prednisone	5.000
Emcompress®	115.930
Explotab®	3.125
Magnesium Stearate	0.945

1 Mix prednisone and **Emcompress®** together for 5 min. Screen through a #20 mesh screen and mix for 15 min.

2 To (1) add **Explotab®** and blend thoroughly for 10 min.
3 Add magnesium stearate to (2), blend for 5 min, and compress.

DoxylamineTablets

(25 mg)

	mg/tablet
Doxylamine Succinate	25.13
Syloid® 244	3.35
Solka-Floc™ BW100	16.70
Emcompress®	331.82
Explotab®	20.00
Magnesium Stearate	3.00

1 Screen doxylamine succinate through #30 mesh screen and blend with **Syloid®** for 10–15 min.
2 To (1) add **Solka-Floc™** and 1/3 **Emcompress®** and mix for 10 min.
3 To (2) add remaining **Emcompress®** and blend thoroughly for 10 min.
4 Add magnesium stearate to (3) and blend 3–5 min.
5 To (4) add **Explotab®**, blend 5–7 min, and compress.

Prednisol Tablet

(5 mg)

	mg/tablet
Prednisolone	5.000
Emcompress®	115.930
Explotab®	3.125
Magnesium Stearate	0.945

1 Mix prednisolone and **Emcompress®** together for 5 min. Screen through a #20 mesh screen and mix for 15 min.
2 To (1) add **Explotab®** and blend thoroughly for 10 min.
3 Add magnesium stearate to (2), blend 5 min, and compress.

Furosemide Tablet

(40 mg)

	mg/tablet
Furosemide	42.00
Polyvinylpyrrolidone	q.s.
Syloid® 244	3.50
Cab-O-Sil®	0.30
Emcompress®	111.00
Explotab®	4.00
Magnesium Stearate	1.20

1 Coat the furosemide by granulating with 4% aqueous solutin of PVP. Pass through #8–10 mesh screen and dry overnight at 45 C. Pass the dried mass through #20 mesh screen.
2 To (1) add **Syloid®** and **Cab-O-Sil®** and mix thoroughly for 10–15 min.
3 Add **Emcompress®** and **Explotab®** to (2) and mix for 10 min.
4 To (3) add magnesium stearate, blend 5–10 min, and compress.

Antihistamine Tablet

(60 mg)

	mg/tablet
Pseudoephedrine HCl	60.00
Chlorpheniramine Maleate	4.00
Emcompress®	78.00
Avicel® PH101	60.00
Explotab®	7.00
Magnesium Stearate	1.00

1 Mix pseudoephedrine, **Emcompress®**, and **Avicel®** together for 10 min.
2 To a small portion of (1) add chlorpheniramine and mix for 10 min.
3 Mix (1) and (2), add **Explotab®** and blend together for 10–15 min
4 To (3) add magnesium stearate, mix 5 min, and compress.

Amitriptyline HCl Tablet

Formula No. 1

(25 mg)

	mg/tablet
Amitriptyline HCl	25.00
Emcompress®	79.78
Explotab®	4.40
Magnesium Stearate	0.82

Mix amitriptyline with **Emcompress®** in geometric dilutions, then add **Explotab®** and mix thoroughly for 15 min. Add magnesium stearate, blend 5–10 min, and compress.

	No. 2	No. 3
	(50 mg)	*(100 mg)*
	mg/tablet	mg/tablet
Amitriptyine HCl	50.00	100.00
Emcompress®	164.32	352.44
Explotab®	9.00	19.00
Magnesium Stearate	1.68	3.56

Mix amitriptyline with **Emcompress®** and **Explotab®** for 15–20 min. Add magnesium stearate, blend 5–10 min, and compress.

Ascorbic Acid Tablet

(250 mg)

	mg/Tablet
Ascorbic Acid	250.00
Veegum® F	30.00
Avicel® PH 101	120.00
Sta-Rx 1500	60.00
Emdex	122.00
Stearic Acid	6.00
Cab-O-Sil® M-7	12.00

A dry lump-free combination of all ingredients except the stearic acid and **Cab-O-Sil®** is blended for 15 min. Add the stearic acid and **Cab-O-Sil®** and blend for an additional 10 min. Compress.

Physical Properties:

Mean Tablet Weight (n = 20) — 586 mg
Mean Hardness — 7.5 kg
Friability — 0.47%
Disintegration Time (purified water) — 2.7 min

Acetaminophen Tablet

Formula No. 1

(325 mg)

	mg/tablet
Acetaminophen	325.00
Veegum® F	30.00
Emdex	230.00
Avicel® PH 101	130.00
Stearic Acid	10.00
Cab-O-Sil®	25.00

Blend all the ingredients, except the stearic acid for 15 min. Add the stearic acid to the mixture and blend for an additional 10 min. Compress into capsule shaped tablets.

Physical Properties:

Mean Tablet Weight (n = 20) — 755 mg
Mean Hardness — 10 kg
Friability — 0.2%
Disintegration Time (purified water) — 3.8 min

No. 2

(325 mg)

	mg/tablet
Acetaminophen USP (granular)	325.00
Solka-Floc™ BW100	120.00
Avicel® PH 101	108.30
Cab-O-Sil®	3.00
Explotab®	14.40
Magnesium Stearate	4.30

1 Mix acetaminophen, **Solka-Floc™**, and **Avicel®** together for 10 min.
2 Add **Cab-O-Sil®** to (1) and mix for 10–15 min.
3 Add **Explotab®** to (2) and blend 10 min.
4 Add magnesium stearate to (3), blend 5 min, and compress at maximum hardness

No. 3

	mg/tablet
Acetaminophen USP (powder)	500.0
Starch 1500™	194.4
Magnesium Stearate USP	1.8

The acetaminophen and **Starch 1500™** are dry-blended in a Hobart mixer. Sufficient ambient temperature water is added, while mixing, to achieve proper massing. (As with most granulations, the quantity of water necessary is variable. An amount equal to twice the weight of the **Starch 1500™** is a good starting point.) The wet massed material is passed through a 10 mesh screen and oven dried at 50 C. The dried granulation is then passed through a 16 mesh screen to achieve final particle size. The magnesium stearate is added and blended in a PK blender.

Tablet Coating

	Suspension	*Dry Solids*
	g	g
Aquacoat	9,667	2,900
Dibutyl Sebacate	580	580
HPMC-E5 (11% sol'n.)	26,364	2,900
Water	8,691	—
Chroma-Kote™ (aq. color)	2,400	2,400
PEG 8000	58	58
Tween® 80	240	240

Physical Properties:
Solids — 19%

Direct Compression Tablet

(5 g, Tablet Press, Kilian Eiffel)

D.C. Sugar **Nu-Tab** }	95
Adjuvants	4
Magnesium Stearate	1

The optimum combination for lubrication of powders consists of:

Compritol 888	1.0
Magnesium Stearate	0.2

Note: A suitable quantity of magnesium stearate limits the problems of capping which may occasionally occur with this excipient.

Antacid

Formula No. 1

(Aluminum Hydroxide-Magnesium Hydroxide Gel)

A	**Al-Mg Fluid Gel**	36.20
B	Sorbitol (70% sol'n.)	7.00
C	**Antifoam AF Emulsion**	1.70

D	Methyl Paraben	0.22
E	Propyl Paraben	0.04
F	Flavor	0.30
G	**Methocel® K4M**	0.60
H	Deionized Water	53.94

Combine C and H using an Eppenbach Homomixer. Add A, mix 10 min. Transfer to overhead stirrer. Combine B and F. Mix thoroughly and add to formula while mixing. Combine D, E, and F. Upon solution add to rest of formula and mix 5 min.

No. 2

(Double-Strength Suspension)

A	**Al-Mg Fluid Gel**	64.40
B	Sorbitol (70% sol'n.)	13.00
C	**Antifoam AF Emulsion**	1.70
D	Methyl Paraben	0.22
E	Propyl Paraben	0.04
F	Flavor	0.30
G	**Methocel® K4M**	0.50
H	Deionized Water	19.84

Combine C and H using an Eppenbach Homomixer. Add A, mix 10 min. Transfer to overhead stirrer. Combine B and F. Mix thoroughly and add to formula while mixing. Combine D, E, and F. Upon solution add to rest of formula and mix 5 min.

No. 3

(Double-Strength Suspension)

A	**Liquigel HO**	25.60
B	Magnesium Hydroxide Paste	22.20
C	Sorbitol (70% sol'n.)	20.00
D	Methyl Paraben	0.22
E	Propyl Paraben	0.04
F	Flavor	0.30
G	**Antifoam AF Emulsion**	1.70

Combine G and H using an Eppenbach Homomixer. Add A, mix 10 min. Add B, mix 20 min. Transfer to overhead stirrer, and add C. Combine D, E, and F. Upon solution add to rest of formula and mix 5 min.

No. 4

(Single-Strength Suspension)

A	**Liquigel HO**	15.20
B	Magnesium Hydroxide Paste	12.00
C	Sorbitol (70% sol'n.)	7.00
D	**Methocel K4M**	0.30
E	Methyl Paraben	0.22
F	Propyl Paraben	0.04
G	**Antifoam AF Emulsion**	1.70
H	Deionized Water	63.24
I	Flavor	0.30

Combine G and H using an Eppenbach Homomixer. Add A, mix 10 min. Add B, mix 20 min. Transfer to overhead stirrer. Combine C and D. Mix thoroughly and add to formula while mixing. Combine E, F, and I. Upon solution add to rest of formula and mix 5 min.

No. 5

(Suspension)

A	**Veegum® HS**	0.5
	Rhodigel 23	0.2
	Water	40.1
B	Sorbitol (70% sol'n.)	20.0
	Aluminum Hydroxide Gel	36.0
	Magnesium Hydroxide USP	3.2
C	Flavor	q.s.
	Preservative	q.s.

Add the dry blend of **Veegum®** and **Rhodigel** to the water slowly, agitating at maximum available shear until smooth. Mix Part B until uniform and add to Part A. Mix until smooth and uniform.

Consistency: Low viscosity.
Suggested Packaging: Bottle.
Note: **Veegum® HS** is used in this formula for its optimum electrolyte stability while effectively suspending the hydroxide acitves. With xanthan gum it provides a nonthixotropic system which does not thicken on storage and remains freely flowable without excessive settling. Sorbitol is added to improve mouth feel and provide sweetening.

Hemmorhoid Salve

Yellow Petrolatum	95
Zinc Undecylenate	5

Burn/Bite Cream

A	**Veegum®**	1.5
	Water	56.8
	Propylene Glycol	3.0
	FG-10 Antifoam	0.2
B	Mineral Oil (light)	10.0
	Acetulan®	7.0
	Menthol (USP)	1.0
	Benzocaine (USP)	5.0
	Syncrowax AW1-C	3.0
	Arlacel® 165	12.0
	Tween® 60	0.5
	Preservative	q.s.

Add the **Veegum®** to the water slowly, agitating at maximum available shear until smooth. Add remaining Part A ingredients and heat to 75–80 C. Heat Part B with stirring, keeping benzocaine suspended, to 75–80 C. Add Part B to A and mix while cooling. Pour off at about 40 C.
Consistency: High viscosity cream.
Suggested Packaging: Tube.
Note: **Veegum®** is used in this formula for high temperature emulsion stability. The oil phase is designed to be bland and emollient with rapid,

nongreasy rub in. Menthol is included for a slight cooling effect. The benzocaine suspended in the internal phase provides topical anesthesia. Combining the **Syncrowax** with the **Arlacel®** emulsifier helps control softening at elevated temperatures.

External Analgesic

	Formula No. 1 *(Lotion)*	No. 2 *(Cream)*
A **Veegum®**	0.75	1.50
Water	72.85	54.70
FG-10 Antifoam	0.10	0.20
Unimulse® C	0.70	—
Propylene Glycol	3.00	3.00
B Methyl Salicylate	15.00	15.00
Menthol	5.00	5.00
Tween® 80	0.60	0.60
Finsolv® TN	2.00	—
Syncrowax AW1-C	—	5.00
Arlacel® 165	—	15.00
C Preservative	q.s.	q.s.

Formula No. 1: Add the **Veegum®** to the water slowly, agitating at maximum available shear until smooth. Mix in additional Part A ingredients in order, allowing each addition to achieve dispersion uniformity. Mix Part B until clear. Add Part B to A, mixing until uniform.

Formula No. 2: Add the **Veegum®** to the water slowly, agitating at maximum available shear until smooth. Mix in additional Part A ingredients in order, allowing each addition to achieve dispersion uniformity. Heat Part A to 75–80 C. Heat Part B while stirring to 75–80 C. Add Part B to A and mix while cooling. Pour off at 40 C.

Note: The formulas spread easily and rub in quickly with rapid heat onset from the methyl salicylate.

Child's Wound Ointment

A	**Emulgade F**	12.00
	Petrolatum	18.00
	Cetiol	6.00
	Mineral Oil	6.00
	Zinc Oxide	10.00
	Talc	10.00
B	Water	37.00
	Germaben® II	1.00

Mix and melt Part A to 70 C. Heat Part B to 70 C and add to Part A. Mix Parts A and B and allow it to cool to room temperature.

"Natural" Hydrocortisone Cream

Oil Phase:

Super Refined Almond Oil	10.0
Crodacol C-95	1.5
Crodesta F-10	2.0
Crodesta F-110	3.5

Water Phase:

Distilled Water	77.0
Propylene Glycol (USP)	4.0
Germaben® II	1.5

Active Phase:

Hydrocortisone Acetate	0.5

Combine oil phase and heat to 78 C with agitation. Combine water phase and heat to 80 C. Slowly add water phase to oil phase under controlled agitation. Allow to cool. When batch temperature reaches 40 C add active phase. Cool to room temperature and fill.

Calamine Lotion

A	**Veegum®**	0.8
	Rhodigel 23	0.2
	Water	79.6
B	Propylene Glycol	3.0
	Tween® 80	0.4
C	Calamine (prepared, USP)	8.0
	Zinc Oxide (USP)	8.0
D	Preservative	q.s.

Add the dry blend of **Veegum®** and **Rhodigel** to the water slowly, agitating at maximum available shear until smooth. Add Part B and mix slowly until smooth. Add Part C slowly and mix until smooth and uniform.

Consistency: Low viscosity lotion.

Suggested Packaging: Bottle.

Note: The lotion dries quickly on the skin and is not readily rubbed off on clothing.

Lesion Ointment

Petrolatum	95
Zinc Undecylenate	5

Acne Cream

Deionized Water	71.0
Glycerin	3.0
Macaloid	2.0
Preservative	0.3
Glyceryl Monostearate (SE)	3.0
Stearyl Alcohol	1.0
Isocetyl Stearate	1.0
Preservative	0.1
Deionized Water	11.6
Benzoyl Peroxide	7.0

Cough Syrup

Formula No. 1

(Tribromoethane)

Bromoform	0.500
Ethanol (80%)	4.500
Glycerin	15.000
Lycasin®	70.400
Demineralized Water	9.600
Methyl Parahydroxybenzoate	0.075
Propyl Parahydroxybenzoate	0.075

Solubilize the methyl and propyl parahydroxybenzoate in ethanol. Heat the demineralized water to 50 C and add to methyl/propyl parahydroxybenzoate solution. Add the bromoform, glycerin, and **Lycasin®**. Agitate until solution is obtained. Filter and bottle.

No. 2

(Codethylene)

Ethylmorphine	0.10
Demineralized Water	0.40
Lycasin®	87.56
Demineralized Water	11.94
Sodium Benzoate	0.20

Solubilize the sodium benzoate in 11.94 parts warm demineralized water. Solubilize the ethyl morphine in 0.40 parts demineralized water and add to sodium benzoate solution. Add above solution to 87.56 parts of **Lycasin®**.

Chapter VIII
METAL TREATMENTS

Alkaline Bright Nickel Plate Bath

	g/l
Sodium Citrate	100
Ammonium Chloride	50
Nickel Chloride	30
Sodium Hypophosphite	10

Adjust to pH 8–10 with ammonia. Maintain pH at 8–10 with ammonia while plating.

Bath temperature: 190 F (room temperature for plastics).

Note: Sodium citrate is used in electroless nickel plating baths where the citrate chelates the nickel and allows the operator to regulate its release onto the substrate which can be metal or plastic. The thickness of the plate, porosity, luster, and hardness are controlled by the citrate.

This plating bath is based on the chemical catalytic reduction of a nickel salt to the metal by sodium hypophosphite. A salt of a precious metal such as gold, platinum, or palladium may be necessary in catalytic amounts to activate the reduction process. This bath will plate at a rate of about 0.3 mil/h at 190 F on metal or at a lower rate at room temperature on plastic.

U.S. Patent 2,532,283.

Tin Plating Bath

Miranol® C2M Anhydrous Acid	0.08–0.12
Sulfuric Acid	4.00–7.00
Stannous Sulfate	2.00–4.00
Nonionic Surfactant	0.40–0.70
Brightener and Antioxidant	q.s.
Water	93.52–88.18

Gold Plate Bath

Formula No. 1

	g/l
Potassium Gold Cyanide	2
Ammonium Chloride	75
Sodium Citrate	50
Sodium Hypophosphite	10

Physical Properties:
pH — 7.0–7.5
Temperature — 198–203 F
Rate — 0.3 mil/h

No. 2

	g/l
Potassium Gold Cyanide	28.00
Citric Acid	60.00
Monopotassium Acid Phthalate	25.00
Tungstic Acid	45.00
Sodium Hydroxide	16.00
N,N Diethylgycine (Na)	3.75

Physical Properties:
pH — 5–6
Temperature — 185–200 F

U.S. Patent 3,123,484

Note for Formula No. 1 and No. 2: Sodium citrate is used to control the thickness, porosity, luster, and hardness of electroless gold plating baths. These systems use chemical reducing agents to reduce the gold cation to gold metal.

Citric/Persulfate Desmutter for Aluminum

Formula No. 1

Citric Acid (anhyd.)	10
Ammonium Persulfate	5
Water	85

Note: The persulfate decomposes easily and must be replenished daily. Use at room temperature with vigorous agitation. Bath life is good to >1% copper concentration in the solution.

No. 2

Citric Acid (anhyd.)	10.0
Ammonium Persulfate	5.0
Ammonium Bifluoride	0.1
Ferric Sulfate	0.5
Water	84.5

Note: This formula is considerably more aggressive than Formula No. 1. The persulfate must be replenished daily. Use at room temperature with vigorous agitation.

Note for Formula No. 1 and No. 2: Citric acid is used in conjunction with persulfate salts for the removal of black smut from aluminum-copper alloys after cleaning with caustic or a nonalkaline cleaner. The persulfate oxidizes the copper to a soluble form. The citric acid then chelates the copper preventing redeposition on the aluminum. These systems do not attack the base metal aluminum.

Citric-Phosphoric Acid Cleaner for Aluminum

Phosphoric Acid (85%)	3
Citric Acid	4
Triton® X-100 or **Tergitol® NPX**	3
Methyl Ethyl Ketone or **Butyl Cellosolve**	4
Water	86

The addition of nitric acid (1%) may improve the brightness.

Note: The addition of a thickening gum such as xanthan will aid in preventing "run off" from vertical surfaces. This formulation will not harm paint and is used on aluminum parts, architectural components and aircraft.

Mono Ammonium Citrate for Derusting Steel

Formula No. 1

Citric Acid (anhyd.)	10
Ammonium Hydroxide (29%)	3
Water	87

Note: Use this mixture at pH 3.5. Above 170 F (76 C) the action is rapid; below that temperature, longer times are required. A corrosion inhibitor can be added.

No. 2

Citric Acid (anhyd.)	5.0
Ammonium Hydroxide (29%)	1.5
Tergitol® 15-S-9	0.2
Water	93.3

Physical Properties:
pH — 3.4

No. 3

Citric Acid (anhyd.)	25
Ammonium Hydroxide (29%)	4
Propylene Glycol	34
Diacetone Alcohol	10
Sodium Tripolyphosphate	1
Water	26

Dissolve citric acid in water, *cautiously* add ammonium hydroxide, then add the remaining ingredients. Dilute 1 part cleaner to 4 parts water for use. A corrosion inhibitor can be added.

Note: This cleaner will attack rust, has good penetrating action, adheres well to the surface and will remove oil and surface grime.

Citric/Persulfate Ferrous Sulfide Scale Dissolver

Citric Acid (anhyd.)	5
Sodium Persulfate	5
Rodine 241	1
Water	89

Recirculate at 185–195 F (85–90 C).

Note: The citric acid chelates the iron, holding it in solution for removal from the equipment. A corrosion inhibitor is necessary to slow the aggressive action of the ferric ion on the base metal.

Calcium Phosphate Scale Remover for Heat Exchangers

Citric Acid	10.0
Armohib 31	0.1
Tergitol® 15-S-9	0.1
Ammonium Bifluoride	0.1–0.5
Water	89.7

Circulate at room temperature or slightly warm until clean. Monitor to be sure the pH is below 2.7.

Notes: Approximately 4 lb of citric acid are required to dissolve 1 lb of calcium phosphate or hydroxyapatite scale.

The use of ammonium bifluoride is not recommended unless necessary to dissolve silicates.

Citric acid is used for removal of calcium phosphate scales deposited on the walls of pipes, pumps, and heat exchangers. Citric acid will dissolve and chelate or sequester calcium and other metal ions only if the pH is held below pH 2.7. The addition of a corrosion inhibitor is recommended. Ammonium bifluoride can be added if the scale contains silicates, although ABF at elevated temperatures can be corrosive.

Citric Acid Radiator Cleaner

Citric Acid (anhyd.)	20.0
Ammonium Hydroxide	18.0
Triton® X405	1.5
Tributyl Phosphate	0.1–0.2
Water	60.6

Application: Use 750 ml of concentrate to treat 1 radiator.

Note: Citric acid neutralized to pH 3.5 with ammonium hydroxide is used to chelate iron oxide scale in automobile cooling systems. It does not readily attack copper or aluminum parts. A surfactant is added to remove grease.

Liquid Rust Remover

Miranol® JEM Conc.	2.0
Potassium Hydroxide (45%)	75.0
Triethanolamine	12.0
Water	11.0

Chapter IX
POLISHES

Auto Wipe and Shine Polish

Formula No. 1

A	**SF96® (100)**	3.50
	Hoechst Wax S	2.60
	Oleic Acid	1.50
	Mineral Spirits	5.00
B	Morpholine	1.40
	Carbopol® 934	0.17
	Triethanolamine	0.17
	Water	85.66

Heat Part A to 90 C (194 F). Heat Part B to 85 C (185 F). Add part A to B with high-shear agitation. When homogeneous, continue mixing with mild agitation until cool.

Application: Clean car surface before application. Pour a small amount on a dry, soft absorbent cloth and apply to a small section of the surface (2′× 2′). Wipe to a shine using a dry section of the cloth for the final shine.

Warning: When solvents are used, proper safety precautions must be observed. All solvents must be considered toxic and must be used only in well ventilated areas. Exposure to high vapor concentration must be avoided. When flammable solvents are used, storage, mixing, and use must be in areas away from heat, sparks, open flames, or other sources of ignition.

No. 2

(Emulsion, Wax-Type)

A	**SF96®** (1000)	5.000
	SR 107	2.000
	Oleic Acid	2.000
	Mineral Spirits	15.000
	Kerosene	15.000
B	Morpholine	1.000
	Carbopol® 934	0.075
	Triethanolamine	0.075
	Water	49.850
C	**Snow Floss**	10.000

1 Mix components of Part A.
2 In Part B disperse the **Carbopol® 934** in water and when completely dissolved, add the triethanolamine. Then add the morpholine just prior to proceeding with Step 3.
3 Add Part A to B with rapid high-shear agitation until a uniform mixture is obtained.
4 Add Part C slowly and continue agitation until a uniform blend is obtained.

Aerosol Packaging: This formula can be packaged as a liquid or used in a foam-type aerosol package. If the latter is desired, use the following procedure:

Check the viscosity of the material with a Brookfield viscometer, Model RVF-7, Spindle No. 4 for 15 s at 20 rpm. The viscosity should be between 3400–4000 cps @ 25 C (77 F).

Aerosol Loading:

Concentrate	94.5
Isobutane	5.5

Application: Remove excess dirt by washing. Apply polish to a clean, soft cloth and apply to the car surface using a circular motion. Allow the polish to dry to a white haze and then buff to a high gloss with a clean, dry cloth.

Warning: When solvents are used, proper safety precautions must be observed. All solvents must be considered toxic and must be used only in well ventilated areas. Exposure to high vapor concentration must be avoided. When flammable solvents are used, storage, mixing, and use must be in areas away from heat, sparks, open flames, or other sources of ignition.

No. 3

(Liquid, O/W Type, Detergent-Resistant)

A	**SF96®** (350)	1.5
	Oleic Acid	2.5
	Kerosene	10.0
	Mineral Spirits	6.0
B	Water	57.6
	Wax Emulsion (20%)	2.5
	Morpholine	2.0
	Carbopol® 934	0.2
	Triethanolamine	0.2
C	Isopropanol	0.4
	Mineral Spirits	5.0
	SF 1710	1.5
	SF 1705	2.1
D	**Kaopolite® SF**	8.5
E	Ammonium Hydroxide	q.s.

Procedure:

1	**Barreco C-36 Wax**	12.0
	Hoechst Wax E	8.0
	Oleic Acid	3.0
	Morpholine	4.0
2	Water	73.0

Melt wax, add oleic acid, and then add morpholine — 87–93 C (189–199 F). Then add preheated water, mix and cool quickly. Disperse **Carbopol® 934** in water. When completely dissolved, add triethanolamine.

Then add the wax emulsion and morpholine. Add Part A to Part B with high-speed agitation. Add Part C. Continue high-speed agitation or mill to a stable emulsion. Add Part D slowly with good agitation. Adjust to pH 9.0 if necessary with ammonium hydroxide.

Application: Apply a thin even coat with a damp cloth using circular motion. Allow the film to dry thoroughly. A little extra time will assure good detergent resistance. Buff with a clean dry cloth turning the cloth frequently to insure maximum ease of buffing.

No. 4

A	**Veegum® T**	1.0
	Water	47.0
B	**Rosswax 63-0552**	1.3
	Rosswax 63-0542	0.2
	Silicone #530 Fluid	0.5
	Silicone #531 Fluid	3.0
	Sorbitan Monostearate	1.0
	Mineral Spirits	36.0
C	**Snow Floss**	up to 10.0
	Preservative	q.s.

No. 5

(Solvent)

A	**G.E. SF 96** (1000)	3.0
	Rosswax 63-0642	1.4
	Rosswax 63-0632	6.0
B	Kerosene (deodorized and decolorized)	50.0
	Amsco Special Naphloite	45.0

Heat Part A to 100 C, using moderate agitation. Heat Part B to 60 C. Add Part B to A with agitation.

Auto Solvent Glaze

A	**SF96® (1000)**	3.0
	Wax OM	1.4
	Wax F	0.6
	SR 107	1.0
B	Mineral Spirits	94.0

Heat the components of Part A to 100 C (212 F) until the waxes are completely melted. Use moderate agitation. Preheat Part B to 60 C (140 F) to prevent precipitation of the wax. Add Part B slowly to Part A with constant agitation.

Application: Spread the film evenly with a clean, dry cloth. Allow the film to dry and then buff with a clean, dry cloth to a high gloss.

Warning: When solvents are used, proper safety precautions must be observed. All solvents must be considered toxic and must be used only in well ventilated areas. Exposure to high vapor concentration must be avoided. When flammable solvents are used, storage, mixing, and use must be in areas away from heat, sparks, open flames, or other sources of ignition.

Furniture Polish

(Solvent Base)

Durmont 500D	12.0
Carnauba Extra Lt. Ref.	4.0
Paraffin 130/135	4.0
Mineral Spirits	40.0
Kerosene	40.0

Melt waxes and heat to 90–95 C. Add hot wax to cold solvent under agitation. Cool under agitation to pour point (2–4° above cloud point).

Floor and Wood Product Polish

(Liquid Solvent)

Duroxon R21	3.0
Alof 0604	2.5
Vestowax FT-150	1.5

Paraffin 150/155	5.0
Mineral Spirit	88.0

Melt the waxes to 110 C. Add the molten wax to the cold mineral spirit. Cool and fill.

Spray and Shine Polish

	Formula No. 1	No. 2
Uniwax 1050 Emulsion (20%)	10.0	—
Carnauba PV 0553 Emulsion (20%)	—	10.0
Silicone Emulsion SM2135	10.0	10.0
Lemon Fragrance	q.s.	q.s.
Water	80.0	80.0

Blend the wax emulsion with the water. Next add the silicone emulsion preblended with the fragrance. Mix all ingredients together.

Household Metal Polish

(Soft Paste)

1	**Hoechst Wax LP**	2
2	**Hoechst Wax E**	3
3	Oleic Acid	4
4	Kerosene (odorless)	20
5	Trichlorethylene	2
6	Ammonium Citrate	4
7	Ammonium Hydroxide (26°)	15
8	Deionized Water	25
9	**Kaopolite® SF**	25

Melt (1) and (2) at 180–190 F, add (3), (4), and (5), reheating to 140 F. In another vessel, dissolve (6) and (7) in (8), heat to 140 F, then add to the previous mixture with stirring. With good agitation, slowly add (9), shut off heat and continue agitation until cooled to below 100 F.

Emulsion Metal Polish

1	Kerosene	12.0
2	VM&P Naphtha	24.0
3	Oleic Acid	2.5
4	Water	43.0
5	Hexametaphosphate	1.0
6	**Kaopolite® SF** or **1168**	15.0
7	Ammonia Water (26°)	1.9
8	Triethanolamine	0.6

Disperse (4), (5), (6), (7), and (8), and add this solution to the oil phase, (1), (2), and (3), agitating well until smooth.

Note: Use **Kaopolite® 1168** for more aggressive cleaning and polishing action.

Silver Polish

Formula No. 1

Miranol® DM	10.0
Ethyl Alcohol	5.0
Snow Floss	7.5
Water	77.5

No. 2

(Tarnish-Retardant, Soft Paste)

1	Water	69.4
2	**Natrosol® 250 LR**	1.0
3	Tarnish Retardant	5.0
4	**Brij® 30**	5.0
5	**Kaopolite® SF**	14.5
6	**Korthix™**	5.0
7	Methyl *p*-Hydroxybenzoate	0.1

Mix (2) into (1). Separately dissolve (3) in (4), with slight warming if necessary, then add to previous mixture. With intensive agitation, add (5), then (6), continuing agitation until well dispersed and smooth. Add (7) as preservative.

No. 3

(Soft Paste)

Miranol® DM	50
Kaopolite® SF	50

Charge 40 parts of **Miranol® DM** to suitable paste mixer. Run mixer continuously while adding **Kaopolite® SF** with further small increments of **Miranol®** if necessary to avoid overloading the mixer drive. The object is to blend at heaviest practicable consistency, to provide thorough dispersion. Continue mixing until fully dispersed and smooth.

Fine Brass Polish

1	Water	39.9
2	Sodium Hexametaphosphate (anhyd.)	0.1
3	**Korthix™**	2.0
4	**Kaopolite® SF**	25.0
5	**Bio-Soft S-100**	3.0
6	Aminoacetic Acid (tech. grade)	5.0
7	Propylene Glycol	25.0
8	Ammonia Water (26°)	q.s.
9	**Colloid 60**	q.s.

Dissolve (2) in (1), add (3) and (4) slowly, with vigorous agitation until well dispersed and smooth. Reduce agitation to moderate, add (5), (6), and (7), blend thoroughly. Continuing moderate agitation, add (8) to pH 8.5. About 0.2% of (9) is effective, if needed, to control foam in mixing and packaging.

Chapter X

ELASTOMERS, POLYMERS, AND RESINS

Styrenic Thermoplastic Elastomers

	Formula No. 1	No. 2	No. 3	No. 4
Kraton® D 1101	100	—	100	—
Solprene 475	—	100	—	100
Antioxidant 330	3	5	3	5
Plastanox ®DLTDP	3	5	3	5
Amoco® Resin 18-290	—	—	10	10
Physical Properties:				
100% Modulus (psi)	360	173	480	247
300% Modulus (psi)	560	239	760	415
600% Modulus (psi)	1620	651	1920	888
Ultimate Tensile	3650	1284	5320	2582
Ultimate Elongation (%)	710	754	790	879

Olefinic Thermoplastic Elastomer

	Formula No. 1	No. 2	No. 3	No. 4
TPR 1800	100	100	—	—
Ren Flex D-58	—	—	100	100
Amoco® Resin 18-290	—	10	—	10

Physical Properties:

Ultimate Tensile Strength	1215	1284	2506	2336
Ultimate Elongation (%)	115	75	425	375
Tear Resistance (lb/in.)	273	266	595	488
Melt Index, 5000 g, 230 C (g/10 min)	0.3	1.6	3.3	6.4

Elastomer Mix

(Energy Saving)

Neoprene W	750
SBR 1614	375
Stearic Acid	5
Scorchguard O	40

Urethane and Isocyanurate Hybrid Rigid Foams

	Formula No. 1	No. 2	No. 3	No. 4
		(Containing DMMP)		
Polymeric Isocyanate	140.1	203.7	160.9	164.7
Sucrose Amine Polyol*	100.0	100.0	50.0	50.0
Terate® 203 Polyol	—	—	50.0	50.0
DMMP	6.7	8.5	7.3	7.5
Dibromoneopentyl Glycol	—	—	—	2.6
Surfactant	2.0	2.0	2.0	2.0
Polycat® 8	1.0	0.5	0.5	0.5
Dabco TMR-2®	—	2.0	2.0	2.0
Refrigerant 11B	40.7	51.6	44.4	45.7

* Typical rigid Sucrose-Amine Polyol with OH number 530 mgKOH/g and 5.7 average functionality.

Physical Properties:				
Composition:				
NCO/OH Index	1.10	1.60	1.60	1.60
DMMP (%)	2.31	2.31	2.30	2.31
Phosphorus (%)	0.58	0.58	0.58	0.58
Bromine (%)	0.00	0.00	0.00	0.48
Foam Density (kg/m³)	30.4	31.0	29.3	28.8
(lb/ft³)	1.90	1.94	1.83	1.80
Foaming Rates (Hand Fixed Foam):				
Stir Time (s)	15	15	15	15
Cream Time (s)	31	39	28	37
Gel Time (s)	120	109	49	94
Rise Time (s)	182	143	56	78
Tack-Free Time (s)	209	160	87	102
Flame Properties by ASTM E-84, 2-in. sample:*				
Max. Flame Spread** (cm)	243	182	168	122
Flame Spread Index	41	31	28	20
Smoke Developed	392	425	287	269
Residual Char Thickness† (cm)	0.0	1.9	0.6	0.8

Unreinforced Casting Compounds

	Formula No. 1	No. 2
	(Soft)	*(Hard)*
Isopolyester Polyol A	100.00	100.00
Diisocyanate	18.50	18.50
Moisture Scavanger	5.00	5.00

---

* Flame tests described above are not intended to reflect hazards presented under actual fire conditions.

** Distance of flame spread beyond the 137 cm igniting flame.

† Thickness of char resulting from decomposition of the 5.1 cm sample situated 30–60 downstream in the tunnel.

Urethane Catalyst	0.04	0.04
Benzoyl Peroxide	—	1.00
Diethylaniline	0.20	0.20
Cobalt Octoate (6%)	0.30	—
Methyl Ethyl Ketone Peroxide	0.30	—
Isocyanate Index	80	80

Unsaturated isopolyester polyol and diisocyanate were thoroughly mixed using a Cowles high-shear mixer. Moisture scavenger and catalysts were added separately. Castings were cured for 16 h at 57 C and post-cured for 2 h at 80 C.

Physical Properties:

Tensile Properties, ASTM D 638		
Strength (psi)	350	11,900
Strength (MPa)	2.4	82
Modulus (10^3 psi)	20	473
Modulus (GPa)	0.14	3.26
Elongation (%)	9.0	5.2
Flexural Properties, ASTM D 790		
Strength (psi)	320	19,200
Strength (MPa)	2.2	132
Modulus (10^3 psi)	20	528
Modulus (GPa)	0.14	3.64
Izod Impact Resistance, ASTM D 256,		
$^1/_8$ in. notched sample (ft-lb/in.)	1.5	1.5
(J/m)	80	80
Durometer Hardness, ASTM D 2240		
Shore D rating	15	73
Glass Transition Temperature (°C)	–23	39

RIM Compound

(Two-Component Hybrid Resin)

A	Diisocyanate	34.70
	Amoco® Styrene Monomer	31.60
	Benzoyl Peroxide	2.60
	Moisture Scavenger	1.70

B	Isopolyester Polyol B	100.00
	Diethylaniline	0.20
	Urethane Catalyst	0.05

A/B Weight Ratio (at 100 isocyanate index) 0.706

In injection rate of 0.5–1.5 lb/s (0.2–0.7 kg/s) and injection pressure of 400–1400 psi (3–10 MPa) worked best. A 3–5 min cycle at 68 C was used although gelation occurred within a few seconds after mixing Parts A and B. Postcuring for up to 2 h at 120 C increased strength and toughness.

Note: A two-component RIM compound keeps the diisocyanate and the unsaturated isopolyester polyol separate and extends pot life. A low molecular weight unsaturated isopolyester polyol needs less styrene diluent. This allows styrene to be added to the diisocyanate side to balance the weight ratio and improve mixing. The low viscosity also allows higher filler and reinforcement levels and reduces the likelihood of "washing" glass mats inlaid in molds.

Hybrid Resin Molding Compounds

	Formula No. 1	No. 2	No. 3
	(Low profile SMC)	*(Highly filled SMC)*	*(Structural SMC)*
Isopolyester Polyol B	55.1	—	—
Isopolyester Polyol C	—	66.4	77.4
Amoco® Polystyrene M9 (40% in styrene)	25.0	20.0	—
Amoco® Resin 18-290 (60% in styrene)	5.0	—	—
Diisocyanate	14.9	13.6	22.6
t-Butyl Perbenzoate	1.5	1.5	1.5
Zinc Stearate	3.0	3.0	3.0
Calcium Carbonate	160.0	250.0	160.0
1″ Chopped Glass Roving (25%)	88.2	118.2	88.2
Isocyanate Index	100	65	100

No. 4

Isopolyester Polyol C	78.0
Isonate 143L	22.0
Baylith L	2.0
Benzoyl Peroxide	1.0
CM-6016	1.5
Isocyanate Index	105

Post-cure was 2 h at 105 C after overnight at room temperature.
Note: Will shatter if severely stressed.

Household Gloves

Natural Rubber Latex e.g., LA-TZ type (60%)	167.0
Potassium Hydroxide (10% sol'n.)	4.0
Potassium Laurate (20% sol'n.)	2.0
Sulfur (50% dispersion)	2.5
Zinc Diethyldithiocarbamate (50% dispersion)	2.0
Antioxidant (50% dispersion)	2.0
Zinc Oxide (50% dispersion)	2.0
Color (dispersion)	as required

The latex mix is matured, strained, and transferred to a dipping tank. Porcelain or aluminum formers at 40–50 C are dipped into a calcium nitrate coagulant and then into the latex. After the required thickness of latex deposit has been obtained, the formers are withdrawn and dried. The gloves are then stripped from the former, leached, and vulcanized: flock may also be applied after dipping. It is customary to reduce the surface friction of the gloves by immersion in dilute chlorine solution for a few seconds, followed by washing and drying.

Rubber Covering for Rollers

	Formula No. 1	No. 2
Natural Rubber (SMR L)*	100.00	100.0
N-660, GPF Black	5.00	—
N-762, SRF-LM-NS Black	—	30.0
Dutrex 729	—	4.0
Zinc Oxide	5.00	5.0
Stearic Acid	1.00	—
Zinc 2-Ethylhexanoate	—	2.0
Wax	1.50	—
2,2,4-Trimethyl-1,2-Dihydroquinoline (polymerized)	2.00	2.0
Zinc 2-Mercaptobenzimidazole	2.00	2.0
Caloxol W5G	3.00	—
Novor 924	2.70	4.2
Sulfur	0.25	0.4
CBS	0.05	—
N-Oxydiethylenebenzothiazole-2-Sulfenamide	—	0.1
TMTM	1.30	—
Tetrabutylthiuram Disulfide	—	1.5
Mix Properties:		
Mooney Scorch, t_5, 120 C (min)	24	16
Curemeter Reversion after 30 min at 180 C (%)	4	10
Vulcanizate Properties:		
Cure Time at 160 C (min)	18	17
Hardness, IRHD	38	49
Tensile Strength (MPa)	23.8	24.7
Elongation at Break (%)	740	605
300% Modulus (MPa)	1.9	5.8
Relaxed Modulus (MPa)	0.65	1.14

* Other high quality latex grades of SMR may also be used.

Compression Set		
70 h at 23 C (%)	8	7
22 h at 70 C (%)	15	18
Resilience, Dunlop Tripsometer (%)	87	85
Tension Fatigue, 0–100% strain, median life (kc)	95	60
After air oven aging for 7 days at 100 C:		
Change in Hardness, IRHD	+1	+2
Change in Tensile Strength (%)	–37	–36
Change in Elongation at Break (%)	–15	–26
Change in 300% Modulus (%)	+21	+39
Tension Fatigue, 0–100% strain, median life (kc)	200	110

Shoe Sole Compounds

Kraton® 4122	100
Styron® 678	0–36
Amoco® Resin 18	0–40
Shellflex® 371 or **Amoco® L-14 Polybutene**	36
Hi-Sil® 233	20
Stabilizers and Inhibitors*	3

Mixer	Banbury B Internal Mixer
Load	1287 g (90% capacity)
rpm	230 (#4 speed)

Order of addition:

1. Dry ingredients and polymer.
2. Add increments of oil as follows:
   $^1/_3$ @ 138 C (280 F)
   $^1/_3$ @ 130 C (265 F)
   $^1/_3$ @ 130 C (265 F)
3. Stop mix at 143–148 C (290–300 F).
4. Cool to about 65 C (150 F) to dump.

---

* One part each: **Irganox® 1010, Tinuvin® P, DLTDP**

Unit Sole for Town Shoe

PVC (K = 65)	100.0
Chemigum P83	20.0
DOP	80.0
Calcium Carbonate	5.0

Stabilizer/Lubricant System:

Epoxidized Soya Bean Oil	3.0
Barium/Zinc	2.5
Calcium Stearate	0.5

Physical Properties:

Hardness (ShoreA) — 60

Notes: Formula exhibits improved flexibility, enhanced abrasion and skid resistance, and good flow characteristics during processing.

Unit Sole for Soccer Boot

PVC (K = 70)	100.0
Chemigum P83	20.0
DOP	65.0

Stabilizer/Lubricant System:

Epoxidized Soya Bean Oil	3.0
Barium/Lead	2.5
Calcium Stearate	0.5

Physical Properties:

Hardness (Shore A)—72

Notes: Formula exhibits improved flexibility, better tear resistance, and good flow characteristics during processing.

Fuel Hose Liner

PVC (K = 65)	100
Chemigum P83	50
DOP	60
Calcium Carbonate	10

Stabilizer/Lubricant System:

Epoxidized Soya Bean Oil	5
Barium/Lead	2

Physical Properties:	*Original*	*1 week in gasoil at R.T.*
Hardness, Shore A	67	67
Tensile (MPa)	16.4	16.4
Tensile (psi)	2425	2425
100% Modulus (MPa)	5.1	5.4
100% Modulus (psi)	750	800
300% Modulus (MPa)	10.5	10.6
300% Modulus (psi)	1550	1575
Elongation at Break (%)	465	375
Weight Loss (%)	—	1.1

Notes: Formula exhibits improved fuel resistance and retained flexibility due to low extractability of plasticizer; better dimensional stability during extrusion; and faster extrusion speeds.

Coextruded Gasket

(Rigid Support/Flexible Seal)

PVC (K = 70)	100
Chemigum P83	40
DIDP	100
Calcium Carbonate	10

Stabilizer/Lubricant System:

Epoxidized Soya Bean Oil	2
Barium/Lead	2

Physical Properties:

Sp. Gr. — 1.17
Hardness (Shore A) — 53
Tensile Strength — 11.1 MPa, 1650 psi
100% Modulus — 2.9 MPa, 425 psi
300% Modulus — 7.1 MPa, 1050 psi
Elongation at Break — 440%
Tear Strength — 135 ppi
Low Temperature Flexibility (Clash & Berg) — –45 C
Compression Set — 60%

Notes: This formula exhibits rubber-like matt finish, excellent low-temperature flexibility and compression set, and faster extrusion rates and improved dimensional stability even from dry blend, giving greater processing latitude even with complex shapes.

Natural Rubber Conveyor Belt Cover

Ribbed Smoke Sheet	100.00
Amoco® L-14 Polybutene or Mineral Oil	4.00
Stearic Acid	2.00
Zinc Oxide	5.00
Carbon Black	45.00
Antioxidants	1.00
Protective Wax	1.00
Sulfur	2.50
Primary Accelerator	1.25
Secondary Accelerator	0.15

Compression molded at 14,000 kPa (2000 psi) into 1.9 mm sheet. Cured at 138 C (280 F): 17.8 min for polybutene compound, 14 min for mineral oil compound.

Automotive Exterior Trim Strip

PVC (K = 70)	100.0
Chemigum P83	25.0
DIDP	55.0

Calcium Carbonate	30.0
Carbon Black	2.0

Stabilizer/Lubricant System:

Epoxidized Soya Bean Oil	4.0
Barium/Lead	2.5

Physical Properties:

Hardness (Shore A)—80

Notes: Formula exhibits fashionable matt finish, good fuel resistance, and faster extrusion speeds and improved dimensional stability even from dry blend.

Injection Molded Automotive Profile

PVC (K = 70)	100.0
Chemigum P83	30.0
DIDP	80.0
Calcium Carbonate	10.0

Stabilizer/Lubricant System:

Epoxidized Soya Bean Oil	3.0
Barium/Lead	3.0
Calcium Stearate	0.5

Physical Properties:

Hardness (Shore A) — 62
Sp. Gr. — 1.2
Tensile Strength — 13.5 MPa, 2000 psi
Low Temperature Flexibility (Clash & Berg) — −43 C
100% Modulus — 4.1 MPa, 600 psi
300% Modulus — 8.8 MPa, 1300 psi
Elongation at Break — 450%
Compression Set — 59%

Notes: Formula exhibits fashionable matt finish, improved compression set, good flow and improved definition on variable sections even from dry blend.

Aluminum Citrate/Polymer

(EOR-Polymer Cross-Linking for the Petroleum Industry)

Inject 25% of the pore volume with anionic partially hydrolyzed polyacrylamide at a concentration of 1500 ppm. Follow with an equal volume of 300 ppm Al^{+++} equivalent to 10,000 ppm 3% aluminum citrate solution. A reservoir could require 1–5 million barrels of injection fluids.

Aluminum citrate can be formulated as a dry blend of aluminum sulfate and sodium citrate or as a convenient solution containing 3% aluminum.

Notes: A quick and convenient performance test for checking the cross-linking of polymers with liquid aluminum citrate has been designed by Miles Laboratories. Copies of this test are available from Citro-tech Products Technical Services Department.

Aluminum citrate is used to gel polymers *in situ* for applications in mobility control and channel blocking in Enhanced Oil Recovery (EOR) operations. The cross-linking (gelling) of the polymer is done by the aluminum ion and is injected in the form of aluminum citrate which has a very slow reaction rate, giving the aluminum a chance to penetrate the polymer. The polymer and aluminum citrate can be pumped down either an injection well or a producer well, depending on the type of treatment desired.

Chapter XI
MISCELLANEOUS

Mold Release Agent

Formula No. 1

Duroxon R-21	6.0
Duroxon E421R	2.0
Duron 0423	2.0
PTFE 727	0.5
Mineral Spirit	89.5

Heat the waxes and 10% of the solvent. Add the remainder of the solvent using rapid agitation.

No. 2

Vestowax A-235	6.0
Petrolatum Ultima	6.0
Silicone Oil (30,000 cps)	8.0
Mineral Spirit	80.0

Heat the wax, petrolatum, and the silicone oil to 110 C. Add 20% of the mineral spirit to the melt. Add the melt to the remainder of the cold mineral spirit using good agitation.

Copper Citrate Algicide

Formula No. 1

(Liquid)

Copper Sulfate	19.8
Citric Acid (anhyd.)	15.2
Water	65.0

This formula contains 5% copper. It can be neutralized to pH 7 if desired using ammonium hydroxide.

Dosage: Heavy growth—1 qt for each 7500 gal or 1000 ft^3 (1.7 ppm Cu). *Lighter growth*—0.4 lb for each 7500 gal or 1000 ft^3 (0.8 ppm Cu).

No. 2

(Dry Blend)

Citric Acid (anhyd.)	100
Copper Sulfate	100

This formula contains 12.5% copper. After blending it is broadcast on the lake surface by hand, by towing a burlap bag of the mix behind a boat, or by using a powered spreader.

Dosage: Heavy growth—0.8 lb for each 7500 gal or 1000 ft^3 (1.7 ppm Cu). *Lighter growth*—0.4 lb for each 7500 gal or 1000 ft^3 (0.8 ppm Cu).

Note for Formulas No. 1 and No. 2: The use of copper citrate for control of algae growth in ponds and lakes has several distinct advantages over copper sulfate treatment.

1 Copper citrate is more soluble in alkaline waters.
2 Copper citrate is less toxic to nontarget organisms.
3 Copper citrate is effective over a longer period of time due to a slow release of copper from the citrate complex.
4 Accidental overdose will not readily lead to toxic levels for other aquatic life.

The higher cost of the copper citrate system is offset by the lower concentrations required, less frequent application, and safety. There will be a slight increase in BOD.

Warning: Treatment of natural waters should be done within the guidelines of local regulations.

Citrate Chelate for Agricultural Micronutrients

Formula No. 1

(5.0% Iron Chelate Sol'n.)

$FeSO_4 \cdot 7H_2O$	24.9
Citric Acid (50%, liquid)	34.7
Water	8.5

Aqua Ammonia (29% NH_3)	to pH 7
Water	q.s. to 100.0

No. 2

(7.0% Copper Chelate Sol'n.)

$CuSO_4 \cdot 5H_2O$	27.6
Citric Acid (50%, liquid)	43.0
Aqua Ammonia (29% NH_3)	to pH 6.5–7.0
Water	q.s. to 100.0

No. 3

(5.0% Manganese Chelate Sol'n.)

$MnSO_4 \cdot 1H_2O$	15.4
Citric Acid (50%, liquid)	36.5
Water	10.0
Aqueous Ammonia (29% NH_3)	to pH 8–9
Water	q.s. to 100.0

Note: Long term stability and compatibility with N-P-K fertilizers can be improved by the addition of 0.1–5% sodium bisulfite as an oxidation inhibitor (Windgasen, 1975). Other antioxidants such as sulfites, dithionites, thiosulfates, and hydroxylamines should also prove effective. Alternatively, the addition of 0.1–2.0%, preferably 0.25% by wt., ferrous ion can be utilized (Hoyt, 1982).

No. 4

(10.0% Zinc Chelate Sol'n.)

ZnO	12.5
Citric Acid (50%, liquid)	60.0
Aqueous Ammonia (29% NH_3)	q.s. to pH 8
Water	q.s. to 100.0

No. 5

(8.0% Zinc Chelate Sol'n.)

ZnO	10.0
Citric Acid (50%, liquid)	48.3
Water	20.0
Aqueous Ammonia (29% NH_3)	q.s. to pH 7
Water	q.s. to 100.0

Note: Should a precipitate begin to appear on long-term storage, the addition of a small amount of aqueous ammonia and stirring brings the precipitate back into solution.

No. 6

(6.0% Cobalt Chelate Sol'n.)

$CoSO_4 \cdot 7H_2O$	29.0
Citric Acid (50%, liquid)	41.0
Aqueous Ammonia (29% NH_3)	q.s. to pH 6
Water	q.s. to 100.0

For Formulas No. 1–6: For each formulation, a reaction vessel capable of thorough mixing should be utilized. Caution should be exercised as the reactions are exothermic in nature.

Add the liquid citric acid and water to the reaction vessel and begin stirring. Slowly distribute the micronutrient source into the vessel being careful to avoid unnecessary clumping and allow to stir for several minutes. Cautiously and slowly add the aqueous ammonia to the mixture until the desired pH is achieved and the solution clears. Dilute to 100% wt/wt with water.

Notes: These formulations can often be adapted to utilize many sources of micronutrients including sulfates, chlorides, and oxides. Note that Formulas No. 1, 2, 5, and 6 remain at or below pH 7. This insures that little or no free ammonia will be given off allowing for easier foliar applications.

Agricultural Sulfur Flowable

(6 lb/gallon)

Veegum®	0.35
Darvan No. 6®	1.50
Water	37.27
Propylene Glycol	7.50
Triton® X-114	0.15
Surfynol® TGE	0.05
Paraformaldehyde	0.10
Flour Sulfur	53.00
Rhodopol® 23	0.08

Add all ingredients, except the **Rhodopol® 23**, to the mill in the order listed. Mix until the sulfur particles are approximately 12–13 μm.* When using a ball mill, about 5 h of grinding time is required to achieve the desired particle size. Add the **Rhodopol® 23** and mill for 15 min or until completely dissolved.

Physical Properties:

This formula was prepared in 2-kg batches in a 1.3-gal ball mill half filled with $1^1/_4$″ Burundum cylinders.

	1 Day	*1 Week*	*1 Month*
Visc. (cps)	480	500	490
Brookfield LV, #3 spindle, 60 rpm			
Stability (1 month):	4 C	20 C	50 C
Gelation	None	None	None
Bleed (% by vol.)	Trace	5%	7%
Packing	None	None	None
Freeze/Thaw (–15 C to 20 C):	No agglomeration, gelation, or bleed through 5 cycles		
Bloom:	Very good, no roping		
Density:	11.7 lb/gal		
pH	7.9		

White Flexoprinting Ink

Rutile Titanium Dioxide	40.00
Versamid 758	15.00
Nitrocellulose E 330	5.00
Lanco-Glidd ET	1.23
Ethyl Acetate	6.00
n-Propanol	14.00
Isopropanol	3.77
Petroleum Solvent SPB 3	15.00

Physical Properties:

Pigment/Resin Ratio — 2:1

Resin Solids — 20%

Eraser

	Formula No. 1	No. 2	No. 3
	(Pencil)	*(Pencil)*	*(Ink)*
Natural Rubber (SMR L)	100	100.0	100.0
White Factice 822	100	400.0	50.0
Maglite D	30	30.0	30.0
Lithopone	100	—	20.0
Barytes	80	—	—
Whiting	150	300.0	—
Pumice	—	50.0	150.0
Titanium Dioxide	10	—	5.0
Fina Process Oil 2059	30	100.0	—
Zinc Oxide	10	10.0	10.0
Stearic Acid	2	2.0	2.0
Sulfur	8	7.0	6.0
MBTS	1	1.5	1.0
DPG	2	2.5	1.5
Pigment		as required	
Mix Properties:			
Mooney Scorch t_5, 120 C (min)	8	60	3

Vulcanizate Properties: Cure: 30 min at 150 C in steam

Density (mg/m^3)	1.67	1.59	1.51
Hardness, IRHD	40	49	67

Mill mixing is often preferred because of the greater control it allows. High factice levels can also cause occasional crumbling in internal mixers unless care is taken. Mill rolls should be kept cool since high temperature may result in sticking. The following mix schedule has been found satisfactory:

Disperse the zinc oxide in the rubber. Add stearic acid, accelerators, sulfur, pigment, filler, and oil. Add the factice.

Erasers are generally either cured from calendered or milled sheet in simple clamped molds in a steam pan or are extruded and then steam cured. Press molding is also used. Molded erasers may be either individually molded or cut to size from vulcanized sheet. The cured erasers are then tumbled with an abrasive to produce a ground surface and to remove any sharp edges. Finally brand marks are stamped on by manually operated or automatic machines.

Crayon Stick

Syncrowax HGL-C	10.0
Paraffin Wax 125/130 F	50.0
Kaolin	25.0
Pigments	15.0

Melt the waxes and heat molten mix to 90 C. Add the kaolin steadily and disperse using a homomixer (high shear mixer). Mix using homomixer for 5 min after adding pigment. Pour into molds about 70 C. Allow to set.

Marking Crayons

Castorwax	18.0
Ceresine Wax	40.0
#1 Yellow Carnauba Wax	17.0
Paraffin Wax 130	20.0
Yellow Beeswax	5.0
Talc	50.0
Pigment	150.0

Melt waxes together at 200 F. Add talc and pigment and pour into molds.

Chain Belt Lubricant

Water	78.0
Monamine ALX-100S	15.0
Monamine R8-26	5.0
Tetrasodium EDTA	2.0

Add ingredients in order listed with good agitation.

Insect Repellent Foam

Polawax® A31	3.0
N,N Diethyl Toluamide	15.0
Ethanol SD-40 (95%)	39.6
Deionized Water	32.4
Propellent 12/114 (40:60)	10.0

German Cockroach Poison

Cab-O-Sil®	0.2
Baygon	1.1
Boric Acid	q.s. to 100.0

Textile Softener

Miranol® DM	71.0
Paraffin Wax	10.0
Tween® 60	1.0
Span 80	1.0
Nonylphenoxypoly(ethyleneoxy) Ethanol	1.0
Water	16.0

Note: This formulation is applied to fabrics such as shirting to enhance needle penetration and sewing speed.

APPENDIX

Federal Laws Regulating Foods, Drugs, and Cosmetics

Anyone who plans to market a food, drug, cosmetic, or chemical specialty should be thoroughly familiar with the Federal laws and regulations pertaining to that particular product.

The Federal Food and Drug Law, which was passed in 1938, was comparatively lenient in its restrictions. But the discovery of more potent drugs, new food additives, and powerful insecticides posed safety problems. In some instances, dangerous toxicity of a new compound was not suspected prior to marketing; then, when its use was widespread, serious poisonings and other deleterious effects were reported.

In 1962, amendments to the 1938 law were passed. These amendments were very restrictive, demanding far more rigid proof of the safety of any drug or chemical used on or by humans; they also brought under Federal control products used on or consumed by both pet and food animals.

Responsibility for enforcement of these laws is still divided among three government agencies. The Food and Drug Administration (FDA), a bureau of the Department of Health, Education, and Welfare, has the broadest regulatory powers. This agency controls the marketing of all foods, medications, and cosmetic products for human use as well as of all veterinary medications.

The Federal Trade Commission (FTC) is charged with the responsibility of policing advertising, to check false, misleading, or illegal statements. One exception is drug advertising to physicians, which comes under FDS jurisdiction.

The Department of Agriculture has jurisdiction over pesticides and their labeling and over additives to foods for animals being raised for human consumption.

Cosmetics

Cosmetics are not restricted as stringently as drugs, but the definition of a cosmetic is very narrow. Any preparation used *only* for cleansing or beautification of the skin, hair, or fingernails is considered a cosmetic. There are certain prohibited drugs or chemicals that are considered too dangerous for use in cosmetics, but on the whole it is up to the manufacturer to market products that are safe. Cosmetic claims can be quite flamboyant, so long as they are only for beautification. Any claim of a medicinal nature, however, even if it is only implied, immediately places the product in the drug class, which means that official permission to market the product is mandatory.

The law does not forbid the use of poisonous substances in a cosmetic. The ban applies only when the quantity of a poisonous ingredient *may* be injurious to users, *under the directions for use as given by the manufacturer.* The product may be extremely hazardous if misused, yet still be acceptable, under the law, *provided* safe directions are included and proper caution statements are made on the label.

While the FTC has charge of all cosmetic *advertising*, it is the FDA that controls *labeling*. The FDA has the authority to seize any product in interstate commerce that it considers misbranded. The term "labeling" is a broad designation; it covers much more than just the label pasted on the container. Any written, printed, or graphic material accompanying the package is considered part of labeling, even if it is delivered by mail, separately from the actual product. A display in a store, along with the product, is also "labeling": so is any literature handed out by demonstrators.

The manufacturer of a cosmetic is thus literally "walking a verbal tight-rope," as the following *partial* list of taboos will further illustrate. Inclusion of any of these designations is sufficient to bring a charge of misbranding a cosmetic:

Circulating cream
Contour cream
Deep pore cleanser
Depilatory for *permanent* removal of hair
Enlarged pore paste
Eyelash grower
Eye wrinkle cream
Hair color restorer
Hair grower
Hair restorer or revitalizing preparation
Muscle oil
Nail grower
Nonallergenic product
Peroxide cream
Scalp food
Skin conditioner or tonic

Skin food or nourishing cream	Wrinkle and double chin eradicator
Skin texture preparation	Any representation that a product is superior in cosmetic value because it contains a vitamin
Stimulating cream	
Tissue cream	

The law does not require that a cosmetic contain a list of its ingredients, but if the labeling puts it in the "drug" class, it must list all ingredients and state the quantities. Any statement that a cosmetic will cure or alleviate dandruff, promote a "healthy skin or scalp," alleviate itching, revitalize dry skin, or cure acne, eczema, freckles, rash, or any "skin trouble" places the product in the "drug" class.

The manufacturer or distributor of a cosmetic has the full responsibility of a truthfully labeled cosmetic, as defined by the FDA. The FDA is not authorized to approve the label or labeling of a cosmetic, but usually will give an informal opinion as to whether a proposed label is to be construed in fact as a drug label. Even such a informal opinion, however, does not relieve the manufacturer of his responsibility, because such an opinion is not an official acceptance of either the product or its labeling.

Additions and amendments to these laws are often made. It is best to write to the proper authority to get up-to-date information.

Some Incompatible Chemicals

The substances in the lefthand column must be stored and handled so that they cannot come into contact with the substances in the righthand column.*

Acetic acid	Chromic acid, nitric acid, hydroxyl-containing compounds, ethylene glycol, perchloric acid, peroxides, permanganates.
Acetone	Concentrate nitric and sulfuric acid mixtures.

* Based partly on *Dangerous Chemicals Code*, 1951 Edition, p. 19–20, Bureau of Fire Prevention, City of Los Angeles Fire Department, published by Parker & Company, Los Angeles 13, California.

Some Incompatible Chemicals *(cont'd.)*

Acetylene	Chlorine, bromine, copper, silver, fluorine, and mercury.
Alkaline and alkaline-earth metals, such as sodium, potassium, cesium, lithium, magnesium, calcium, aluminum	Carbon dioxide, carbon tetrachloride, and other chlorinated hydrocarbons. (Also prohibit water, foam, and dry chemical on fires involving these metals.)
Ammonia (anhydr.)	Mercury, chlorine, calcium hypochlorite, iodine, bromine, and hydrogen fluoride.
Ammonium nitrate	Acids, metal powders, flammable liquids, chlorates, nitrites, sulfur, finely divided organics or combustibles.
Aniline	Nitric acid, hydrogen peroxide.
Bromine	Ammonia, acetylene, butadiene, butane and other petroleum gases, sodium carbide, turpentine, benzene, and finely divided metals.
Calcium carbide	Water (See also acetylene).
Calcium oxide	Water.
Carbon, activated	Calcium hypochlorite.
Chlorates	Ammonium salts, acids, metal powders, sulfur, finely divided organics or combustibles.
Chlorine	Ammonia, acetylene, butadiene, butane and other petroleum gases, hydrogen, sodium carbide, turpentine, benzene, and finely divided metals.
Chlorine dioxide	Ammonia, methane, phosphine, and hydrogen sulfide.
Chromic acid	Acetic acid, naphthalene, camphor, glycerol, turpentine, alcohol, and other flammable liquids.

Some Incompatible Chemicals *(cont'd.)*

Copper	Acetylene, hydrogen peroxide.
Fluorine	Isolate from everything.
Hydrocarbons (benzene, butane, propane, gasoline, turpentine, etc.)	Fluorine, chlorine, bromine, chromic acid, sodium peroxide.
Hydrocyanic acid	Nitric acid, alkalis.
Hydrofluoric acid, anhydrous (hydrogen fluoride)	Aqueous or anhydrous ammonia.
Hydrogen peroxide	Copper, chromium, iron, most metals or their salts, any flammable liquid, combustible aniline, nitro-methane.
Hydrogen sulfide	Fuming nitric acid, oxidizing gases.
Iodine	Acetylene, anhydrous or aqueous ammonia.
Mercury	Acetylene, fulminic acid, ammonia.
Nitric acid (conc.)	Acetic acid, aniline, chromic acid, hydrocyanic acid, hydrogen sulfide, flammable liquids, flammable gases, and nitritable substances.
Nitroparaffins	Inorganic bases.
Oxalic acid	Silver, mercury.
Oxygen	Oils, grease, hydrogen, flammable liquids, solids or gases.
Perchloric acid	Acetic anhydride, bismuth and its alloys, alcohol, paper, wood, grease, oils.
Peroxides, organic	Organic or mineral acids; avoid friction.
Phosphorus (white)	Air, oxygen.
Potassium chlorate	Acids (See also chlorate.)
Potassium perchlorates	Acids (See also perchloric acid.)
Potassium permanganate	Glycerol, ethylene glycol, benzaldehyde, sulfuric acid.

Some Incompatible Chemicals *(cont'd.)*

Silver	Acetylene, oxalic acid, tartaric acid, fulminic acid, ammonium compounds.
Sodium	See alkaline metals.
Sodium nitrate	Ammonium nitrate and other ammonium salts.
Sodium oxide	Water.
Sodium peroxide	Any oxidizable substances, such as ethanol, methanol, glacial acetic acid, acetic anhydride, benzaldehyde, carbon disulfide, glycerol, ethylene glycol, ethyl acetate, methyl acetate, and furfural.
Sulfuric acid	Chlorates, perchlorates, permanganates.
Zirconium	Prohibit water, carbon tetrachloride, foam, and dry chemical on zirconium fires.

Tables

Weights and Measures
Troy Weight

24 gr	=	1 pwt
20 pwt	=	1 oz
12 oz	=	1 lb

Apothecaries' Weight

20 gr	=	1 scruple
3 scruples	=	1 dr
8 dr	=	1 oz
12 oz	=	1 lb

The ounce and pound are the same as in Troy Weight.

Avoirdupois Weight

$27^{11}/_{32}$ gr	=	1 dr
16 dr	=	1 oz
16 oz	=	1 lb
2000 lb	=	1 short ton
2240 lb	=	1 long ton

Dry Measure

2 pt	=	1 qt
8 qt	=	1 pk
4 pk	=	1 bu
36 bu	=	1 chaldron

Tables *(cont'd.)*

Liquid Measure

4 gi	=	1 pt
2 pt	=	1 qt
4 qt	=	1 gal
$31\frac{1}{2}$ gal	=	1 bbl
2 barrels	=	1 hogshead
1 tsp	=	$\frac{1}{6}$ oz
1 tblsp	=	$\frac{1}{2}$ oz
16 fl oz	=	1 pt

Square Measure

144 $in.^2$	=	1 ft^2
9 ft^2	=	1 yd^2
$30\frac{1}{4}$ yd^2	=	1 rod^2
43,560 ft^2	=	1 acre
40 rod^2	=	1 rood
4 roods	=	1 acre
640 acres	=	1 $mile^2$

METRIC EQUIVALENTS
Length

1 in.	=	2.54 cm
1 ft	=	0.305 m
1 yd	=	0.914 m
1 mile	=	1.609 km
1 cm	=	0.394 in.
1 m	=	3.281 ft
1 m	=	1.094 yd
1 km	=	0.621 mile

Capacity

1 U.S. fl oz	=	29.573 ml
1 U.S. liquid qt	=	0.946 l
1 U.S. dry qt	=	1.101 l
1 U.S. gal	=	3.785 l
1 U.S. bu	=	0.3524 hl
1 $in.^3$	=	16.4 m^3

Capacity *(cont'd.)*

1 m	=	0.034 U.S. fl oz
1 l	=	1.057 U.S. liq. qt
1 l	=	0.908 U.S. dry qt
1 l	=	0.264 U.S. gal
1 hl	=	2.838 U.S. bu.
1 cm^3	=	0.061 $in.^3$
1 l	=	1000 ml
1 l	=	1000 cm^3

Circular Measure

60 s	=	1 min
60 min	=	1 degree
360 degrees	=	1 circle

Long Measure

12 in.	=	1 ft
3 ft	=	1 yd
$5\frac{1}{2}$ yd	=	1 rod
5280 ft	=	1 stat. mile
320 rods	=	1 stat. mile

Weight

1 gr	=	0.065 g
1 apoth. scruple	=	1.296 g
1 av oz	=	28.350 g
1 troy oz	=	31.103 g
1 av. lb	=	0.454 km
1 troy lb	=	0.373 km
1 g	=	15.432 gr
1 g	=	0.772 apoth. scruple
1 g	=	0.035 av oz
1 g	=	0.032 troy oz
1 km	=	2.205 av lb
1 km	=	2.679 troy lb

Atomic Weights

(Alphabetical Order)

Element	*Symbol*	*Atomic Number*	*Atomic Weight**
Actinium	Ac	89	(227)
Aluminum	Al	13	26.9815
Americium	Am	95	(243)
Antimony	Sb	51	121.75
Argon	Ar	18	39.948
Arsenic	As	33	74.9216
Astatine	At	85	(210)
Barium	Ba	56	137.34
Berkelium	Bk	97	(247)
Beryllium	Be	4	9.0122
Bismuth	Bi	83	208.980
Boron	B	5	10.811
Bromine	Br	35	79.909
Cadmium	Cd	48	112.40
Calcium	Ca	20	40.08
Californium	Cf	98	(251)
Carbon	C	6	12.01115
Cerium	Ce	58	140.12
Cesium	Cs	55	132.905
Chlorine	Cl	17	35.453
Chromium	Cr	24	51.996
Cobalt	Co	27	58.9332
Copper	Cu	29	63.54
Curium	Cm	96	(247)
Dysprosium	Dy	66	162.50
Einsteinium	Es	99	(254)
Element 102		102	(254)
Erbium	Er	68	167.26
Europium	Eu	63	151.96
Fermium	Fm	100	(253)
Fluorine	F	9	18.9984
Francium	Fr	87	(223)

* A value given in parentheses denotes the mass number of the longest-lived isotope.

Element	*Symbol*	*Atomic Number*	*Atomic Weight**
Gadolinium	Gd	64	157.25
Gallium	Ga	31	69.72
Germanium	Ge	32	72.59
Gold	Au	79	196.967
Hafnium	Hf	72	178.49
Helium	He	2	4.0026
Holmium	Ho	67	164.930
Hydrogen	H	1	1.00797
Indium	In	49	114.82
Iodine	I	53	126.9044
Iridium	Ir	77	192.2
Iron	Fe	26	55.847
Krypton	Kr	36	83.80
Lanthanum	La	57	138.91
Lawrencium	Lw	103	(256)
Lead	Pb	82	207.19
Lithium	Li	3	6.939
Lutetium	Lu	71	174.97
Magnesium	Mg	12	24.312
Manganese	Mn	25	54.9380
Mendelevium	Md	101	(256)
Mercury	Hg	80	200.59
Molybdenum	Mo	42	95.94
Neodymium	Nd	60	144.24
Neon	Ne	10	20.183
Neptunium	Np	93	(237)
Nickel	Ni	28	58.71
Niobium	Nb	41	92.906
Nitrogen	N	7	14.0067
Osmium	Os	76	190.2
Oxygen	O	8	15.9994
Palladium	Pd	46	106.4
Phosphorus	P	15	30.9738
Platinum	Pt	78	195.09
Plutonium	Pu	94	(244)
Polonium	Po	84	(209)

* A value given in parentheses denotes the mass number of the longest-lived isotope.

Element	*Symbol*	*Atomic Number*	*Atomic Weight**
Potassium	K	19	39.102
Praseodymium	Pr	59	140.907
Promethium	Pm	61	(145)
Protactinium	Pa	91	(231)
Radium	Ra	88	(226)
Radon	Rn	86	(222)
Rhenium	Re	75	186.2
Rhodium	Rh	45	102.905
Rubidium	Rb	37	85.47
Ruthenium	Ru	44	101.07
Samarium	Sm	62	150.35
Scandium	Sc	21	44.956
Selenium	Se	34	78.96
Silicon	Si	14	28.086
Silver	Ag	47	107.870
Sodium	Na	11	22.9898
Strontium	Sr	38	87.62
Sulfur	S	16	32.064
Tantalum	Ta	73	180.948
Technetium	Tc	43	(97)
Tellurium	Te	52	127.60
Terbium	Tb	65	158.924
Thallium	Tl	81	204.37
Thorium	Th	90	232.038
Thulium	Tm	69	168.934
Tin	Sn	50	118.69
Titanium	Ti	22	47.90
Tungsten	W	74	183.85
Uranium	U	92	238.08
Vanadium	V	23	50.942
Xenon	Xe	54	131.30
Ytterbium	Yb	70	173.04
Yttrium	Y	39	88.905
Zinc	Zn	30	65.37
Zirconium	Zr	40	91.22

* A value given in parentheses denotes the mass number of the longest-lived isotope.

Emergency First Aid for Chemical Injuries

CHEMICAL	SKIN CONTACT	SWALLOWING
ACIDS Acetic Hydrochloric Phosphoric Sulfuric	Immediately wash with plenty of running water. Remove contaminated clothing. Once all contact areas have been thoroughly washed, apply a mild alkaline solution such as soda bicarbonate. If eyes are involved, immediately flush with warm water for at least 15 min.	Do not induce vomiting. Do not give anything to an unconscious person by mouth. If conscious, wash out mouth with plenty of water, then give milk mixed with egg whites. If this is not available, give as much water as possible. Call doctor.
Hydrofluoric	Wash thoroughly and apply a magnesium oxide paste immediately. Call doctor.	Do not induce vomiting. Give plenty of water. Keep victim prone; do not move.
Chromic Acid & Dichromates	Wash with 5% sodium thiosulfate. If lesions appear, call doctor. Petrolatum in the nasal passages will protect nose from brief exposure.	Hospitalize immediately. Give soda bicarbonate solution immediately. Keep victim warm. Call doctor.
ALKALIS Ammonium Hydroxide Sodium Hydroxide Potassium Hydroxide	Wash with large quantities of water and neutralize with vinegar.	Give plenty of water mixed with lemon juice or vinegar. Follow with a spoonful of salad oil. Call doctor.
CYANIDE SALTS	Wash thoroughly. If lesions appear, call doctor.	Induce vomiting immediately. Give hydrogen peroxide mixed with water. Call doctor.

Emergency First Aid for Chemical Injuries *(cont'd.)*

CHEMICAL	SKIN CONTACT	SWALLOWING
CHLORIDE COMPOUNDS		
Ammonium Chloride Cobalt Chloride Iron Chloride	Wash thoroughly.	Induce vomiting. Give large amounts of water. Use Epsom salts as a laxative. Call doctor.
Antimony Chloride Nickel Chloride Tin Chloride	Wash thoroughly. Treat with lanolin ointment. Call doctor if irritation continues.	Give large quantities of water. Call doctor.
Cadmium Chloride		Give large quantities of water. Call doctor.
NITRATES		
Potassium Nitrate Mercuric Nitrate	Wash thoroughly. Call doctor if rash develops.	Immediately give plenty of water mixed with sodium bicarbonate, then give nonfat milk mixed with raw eggs. Call doctor.
Silver Nitrrate	Wash with salt water and treat for burns.	Give 3 tablespoons of salt mixed in a pint of water. Induce vomiting. Call doctor.
SULFATES		
Aluminum, Ammonium, Cobalt, Copper, Magnesium, Nickel, Potassium, Sodium, Zinc	Wash thoroughly. Call doctor if skin reaction occurs.	Give plenty of water. Call doctor if any reactions occur.
Cadmium Sulfate	Wash thoroughly. Call doctor if skin reaction occurs.	Give plenty of water. Call doctor.

TRADEMARK CHEMICALS

Where to Buy Them

Numbers to the right of each item refer to suppliers who are given in the list of sellers directly after this list.

Chemicals not sold under a trademark may be located in the annual *Buyers' Guide* published by Chemical Week, 330 W. 42 St., New York, N.Y. 10036, and in the Green Book published by the *Oil, Paint and Drug Reporter*, 100 Church St., New York, N.Y. 10007.

TRADEMARK CHEMICALS SUPPLIERS

1	Abbott Laboratories	N. Chicago, IL
2	Air Products & Chemicals	Allentown, PA
3	Ajinomoto USA Inc.	New York, NY
4	Alcolac Inc.	Baltimore, MD
5	Allied-Signal Inc.	Morristown, NJ
6	Alpine Aromatics International Inc.	Metuchen, NJ
7	Alzo Inc.	Matawan, NJ
8	Amchem Products Inc.	Ambler, PA
9	Amerchol Corp.	Edison, NJ
10	American Cyanamid Co.	Wayne, NJ
11	American Hoechst Corp.	Somerville, NJ
12	American Maize Products Co.	New York, NY
13	Amoco Chemicals Cor.	Chicago, IL
14	Anchor Chemical Ltd.	Manchester UK
15	Angus Chemical Co.	Northbrook, IL
16	Arco Chemical Co.	Philadelphia, PA
17	Armak Industrial Chemical Div.	Chicago, IL
18	Asarco Inc.	New York, NY
19	Ashland Chemical Co.	Columbus, OH
20	BASF Wyandotte Corp.	Parsippany, NJ
21	Beatrice Foods Co.	Chicago, IL
22	Buckman Laboratories, Inc.	Memphis, TN
23	Burgess Pigment Co.	Macon, GA
24	Byk Mallinckrodt	Rahway, NJ
25	Cabot	Tuscola, IL
26	Cargill Inc.	Longview, TX
27	CasChem Inc.	Bayonne, NJ
28	Celanese Corp.	Louisville, KY
29	Ciba-Geigy Corp.	Ardsley, NY
30	Clark Colors Inc.	South Plainfield, NJ
31	Colloids Inc.	Newark, NJ

32	Color Techniques Inc.	New York, NY
33	Columbian Div. of City Service Co.	Atlanta, GA
34	Cosan Chemical Corp.	Clifton, NJ
35	Costec Inc.	Palatine, IL
36	Creations Aromatiques	New York, NY
37	Creative Perfumers & Flavorists, Inc.	New York, NY
38	Croda Chemical Ltd.	North Humberside, UK
39	Croda Inc.	New York, NY
40	Croda Surfactants Inc.	New York, NY
41	Croda Universal Ltd.	Hull, UK
42	Crompton & Knowles Corp.	Fair Lawn, NJ
43	Crowley Chemical Co., Inc.	New York, NY
44	Cyclo Chemical Corp.	Miami, FL
45	Cyprus Industrial Minerals Co.	Los Angeles, CA
46	Desert Minerals Inc.	Los Angeles, CA
47	Diamond Shamrock Corp.	Morristown, NJ
48	Diehl, Inc.	New York, NY
49	Domtar Inc.	Mississauga, Ontario, Canada
50	Dow Chemical Co.	Midland, MI
51	Dow Corning Corp.	Midland, MI
52	Dow Corning Ophthalmics, Inc.	Midland, MI
53	Drew Chemical Corp.	Boonton, NJ
54	Du Pont de Nemours, E.I., & Co.	Wilmington, DE
55	Durachem/Dura Commodifies Corp.	Harrison, NY
56	DuraWax	McHenry, IL
57	Durham Chemical Ltd.	Birtley, Co. Durham, UK
58	Durkee Ind. Foods, SCM Corp.	Cleveland, OH
59	Duveen Soap Co.	Brooklyn, NY
60	Eastman Chemical Products Co.	Kingsport, TN
61	Elkem Chemicals Inc.	Pittsburgh, PA
62	Emery Industries Inc.	Linden, NJ
63	Engelhard Minerals & Chemicals Corp.	Edison, NJ
64	Ethyl Corp.	Baton Rouge, LA
65	Evans Chemetics/W.R. Grace & Co.	Darien, CT
66	Exxon Chemical Americas	Houston, TX
67	Fidco/Div. of The Nestle Co.	White Plains, NY
68	Finetex Inc.	Elmwood Park, NJ
69	Florasynth Labs Inc.	New York, NY
70	FMC Corp., Marine Colloids Div.	Philadelphia, PA
71	Freeman Chemical Corp.	Pt. Washington, WI

72	GAF Corp.	Wayne, NJ
73	Gattefose Corp.	Elmsford, NY
74	Gattefossé Ets.	Boulogne, France
75	General Electric Co.	Waterford, NY
76	General Mills Corp. Chemical Div.	Minneapolis, MN
77	Georgia Kaolin Co.	Elizabeth, NJ
78	Gist-Brocades USA	Charlotte, NC
79	Givaudan Corp.	Clifton, NJ
80	Glidden-Durkee, Div. SCM Corp.	Jacksonville, FL
81	Glyco Chemical Inc.	Greenwich, CT
82	Goodrich Co., B.F.	Cleveland, OH
83	Goodyear Tire & Rubber Co.	Akron, OH
84	Grace & Co., W.R./Hampshire Chem.	Lexington, MA
85	Grace & Co., W.R./Organic Chem. Div.	Lexington, MA
86	Grain Processing Corp.	Muscatine, IA
87	Halox Pigments Div., Hammond Lead	Hammond, IN
88	Hampshire Chemical Co.	Lexington, MA
89	Harwick Chemical Corp.	Akron, OH
90	Henkel Chem. Canada Ltd.	Ville d'Anjou, Quebec, Canada
91	Henkel Corp. Process Chemical Div.	Morristown, NJ
92	Henkel Corp. Resin Div.	Minneapolis, MN
93	Henkel Inc.	Teaneck, NJ
94	Hercules Inc.	Wilmington, DE
95	Heterene Chemical Co. Inc.	Paterson, NJ
96	Heubach	New York, NY
97	Hoechst UK Ltd.	Middlesex, UK
98	Hoffman-La Roche Inc.	Nutley, NJ
99	Hormel	Cedar Knolls, NJ
100	Huber Corp., J.M.	Havre De Grace, MD
101	Huls America Inc./Dynamit Nobel AG	Rockleigh, NJ
102	ICI Americas Inc.	Wilmington, DE
103	ICI Americas Inc./Atlas Div.	Wilmington, DE
104	Illinois Minerals Co.	Cairo, IL
105	Indusmin Ltd.	Toronto, Canada
106	Inolex Corp.	Philadelphia, PA
107	International Talc Co.	New York, NY
108	Interstab Chemicals Inc.	New Brunswick, NJ
109	Jefferson Chemical Co.	Austin, TX
110	Johns-Manville Prod. Co.	Denver, CO
111	Jordan Chemical Co.	Folcroft, PA

112	Kaopolite Inc.	Union NJ
113	Keil Chemical Div.	Hammond, IN
114	Kelco Co./Div. Merck & Co. Inc.	San Diego, CA
115	Kerr-McGee Chemical Corp.	Oklahoma City, OK
116	King Industries	Norwalk, CT
117	Lanaetex Products Inc.	Elizabeth, NJ
118	Land O' Lakes	Minneapolis, MN
119	Langer & Co.	Bremen, Germany
120	Lextar	New York, NY
121	Lipo Chemicals Inc.	Paterson, NJ
122	Lonza Inc.	Fair Lawn, NJ
123	Lowenstein Dyes & Cosmetics, Inc.	Brooklyn, NY
124	3M Co.	St. Paul, MN
125	Madis Laboratories Inc., Dr.	South Hackensack, NJ
126	Maggi	New York, NY
127	Mallinckrodt Inc.	St. Louis, MO
128	Malmstrom Chemical Corp.	Linden, NJ
129	Maybrook Inc.	Lawrence, MA
130	Mazer Chemicals Inc.	Gurnee, IL
131	McIntyre Chemical Co. Ltd.	Chicago, IL
132	Mearl Corp.	New York, NY
133	Meer Corp.	North Bergen, NJ
134	Merck & Co.	Rahway, NJ
135	Merck AG, E.	Darmstadt, W. Germany
136	Mineral Pigments Corp.	Beltsville, MD
137	MiniFIBERS Inc.	Johnson City, TN
138	Miranol Chemical Co., Inc.	Dayton, NJ
139	Mississippi Lime Co.	Ste. Genevieve, MO
140	Mobay Chemical Corp.	Pittsburgh, PA
141	Mona Industries Inc.	Paterson, NJ
142	Monsanto Co.	St. Louis, MO
143	Morton Chemicals	Woodstock, IL
144	Nalco Chemical Co.	Naperville, IL
145	National Starch & Chemical Corp.	Bridgewater, NJ
146	Neville Chemical Co.	Pittsburgh, PA
147	New Jersey Zinc Co. Inc.	Palmerton, PA
148	NL Industries	Hightstown, NJ
149	Novarome Inc.	Fairfield, NJ
150	Nuodex Inc.	Piscataway, NJ
151	NYCO	Willsboro, NY

152	Old Hickory Products	New York, NY
153	Olin Chemicals	Stamford, CT
154	Omya Inc.	Proctor, VT
155	Onyx Chemical Co.	Jersey City, NJ
156	Patco Products	Kansas City, MO
157	Patterson, C.J.	Kansas City, MO
158	Penreco Div	Butler, PA
159	Perry Bros.	Woodside, NY
160	Petrofina	New York, NY
161	Pfizer Inc.	New York, NY
162	Philadelphia Quartz Co.	Valley Forge, PA
163	Phillips Chemical Co./Phillips Petroleum	Bartlesville, OK
164	Plasticolors, Inc.	Ashtabula, OH
165	Polyesther Corp.	Southampton, NY
166	PPG Industries	Pittsburgh, PA
167	PQ Corp.	Valley Forge, PA
168	Procter & Gamble Co.	Cincinnati, OH
169	PVO International, Inc.	Boonton, NJ
170	Reheis Chemical Co.	Berkeley Heights, NJ
171	Ren Flex, Inc.	New York, NY
172	Rhone Poulenc Industries	Paris, France
173	Richardson Co., The	Melrose Park, IL
174	Richter, Dr. Kurt GmbH	Berlin, West Germany
175	Ridge Tech Foods	La Crosse, WI
176	RITA Corp.	Crystal Lake, IL
177	Robertet Inc., P.	Oakland, NJ
178	Rohm & Haas Co.	Philadelphia, PA
179	Rona Pearl	Bayonne, NJ
180	Roquette (UK) Ltd.	Kent, UK
181	Ross, Frank B. Co., Inc.	Jersey City, NJ
182	Roure Bertrand Dupont Inc.	Teaneck, NJ
183	Sandoz Chemical Corp.	Charlotte, NC
184	Shell Chemical Co.	Houston, TX
185	Sherex Chemical Co.	Dublin, OH
186	Sherwin Williams Chemicals	Cleveland, OH
187	Smith Chemical & Color Co., Inc.	Jamaica, NY
188	Spencer Kellogg	Buffalo, NY
189	Staley Mfg. Co., A.E.	Decatur, IL
190	Stange	New York, NY
191	Stauffer Chemical Co.	Westport, CT

192	Stepan Co.	Northfield, IL
193	Sturge, John and E., Ltd.	Birmingham, UK
194	Sun Chemical Corp./Pigments Div.	Cincinnati, OH
195	Sutton Laboratories Inc.	Chatham, NJ
196	SWS Silicones	Adrian, MI
197	Synfleur Scientific Laboratories Co.	New York, NY
198	Synthetic Prod. Co.,Div. of Dart Industries Inc.	Stratford, CT
199	Takasaco Corp. USA	Long Island City, NY
200	Tenneco Chemicals Inc.	Piscataway, NJ
201	Textilana Corp.	Hawthorne, CA
202	Thompson, Weinman & Co.	Cartersville, GA
203	Troy Chemical Corp. Inc.	Newark, NJ
204	Tulco Inc.	North Billerica, MA
205	Ultra Adhesives Inc.	Paterson, NJ
206	Ungerer & Co.	New York, NY
207	Union Carbide Corp.	Danbury, CT
208	Union Oil	San Francisco, CA
209	Uniroyal Inc.	Naugatuck, CT
210	U.S. Emulsifiers	New York, NY
211	USI Chemicals Co.	Cincinnati, OH
212	Van Dyk & Co., Inc.	Belleville, NJ
213	Vanderbilt, R.T. & Co., Inc.	Norwalk, CT
214	Westvaco Chemical Div.	Charleston Heights, SC
215	Whittaker, Clark & Daniels Inc.	South Plainfield, NJ
216	Wickhen Products Inc.	Huguenot, NY
217	Witco Chemical Co.	New York, NY
218	Wolff, Walsrode	Walsrode, W. Germany
219	Wyrough & Loser Inc.	Trenton, NJ
220	Zumbro	New York, NY

INDEX

www.ingramcontent.com/pod-product-compliance
Ingram Content Group UK Ltd.
Pitfield, Milton Keynes, MK11 3LW, UK
UKHW041132071026
15587UKWH00009B/438

9 780820 603247